CLEAN AND DECENT

CLEAN AND DECENT

THE FASCINATING HISTORY OF THE BATHROOM & THE WATER CLOSET

and of sundry
Habits, Fashions & Accessories of
THE TOILET
principally in
**GREAT BRITAIN, FRANCE,
& AMERICA**

by
LAWRENCE WRIGHT
M.A., B.ARCH., A.R.I.B.A.

Published by Routledge & Kegan Paul
LONDON

First published 1960
by Routledge & Kegan Paul Limited
Broadway House, Carter Lane, E.C.4

Printed in Great Britain
by Butler & Tanner Limited
Frome and London

© *Lawrence Wright 1960*

Second Impression 1960
Third Impression 1960

To

MOLLY A. MONTGOMERY

*who first turned on
the tap*

Contents

Author's Preface

LIFE never works out according to plan, and a common experience is that of suddenly becoming aware of the extreme oddness of what one is doing. An ordinary citizen caught up in a war finds himself riding a bicycle round Mount Etna, and asks himself in a flash of sanity what on earth he is at. It is with that feeling of 'How ever did *I* get involved in this?' that I find myself finishing a book about baths and water-closets. This is not my field: I am neither plumber nor social historian. I seem to have been trapped into a slightly ridiculous situation, by easy stages. As Thurber says, *things close in.*

Mrs. Montgomery, who runs the Building Exhibition at Olympia, began it all. In a rash moment I agreed to organise and design a 'feature' for this famous show, before a subject had been found. A 'feature', it should be explained, is not just an exhibition stand, and is not tied up with any particular manufacturer; it is a display on a general theme, provided by the exhibition organisers for the entertainment and the incidental instruction of their public. The subject could be something like *The Kitchen Beautiful* or *Whither Stonemasonry?* or *The Larder Through the Ages.* It was Mrs. Montgomery who hit on *The History of the Bathroom* for that particular year, and I found myself exploring some peculiar ground. As my researches progressed, the subject began to prove more interesting than I had expected, and the material fascinating, if all too plentiful. Surprising things came to light. Who would have supposed that the Romans had lagged hot-water cylinders—that Queen Elizabeth I had a valve water-closet—that Louis XIV had cushions in his bath—that baths have been concealed in sofas, and wash-basins in pianos—that whiskey may be added to the bathwater but that mutton chops should not be eaten in the bath—that the shower-bath calls for a hat, and can cause asphyxia—that sponges have sex?

For the exhibition, baths came from all over England: a boot-bath from Devon, a shower-spray-plunge bath from Macclesfield, a hooded bath from Leeds. Wash-basins came, and pan-closets, lovely examples of the potter's art, so skilfully patterned that their modern counterparts seemed very dull. Queen Victoria's railway wash-basins came; wooden water pipes, a musical chamber-pot, Roman razors and the first of all geysers. As all these things fell into their places as milestones of social history, it seemed that more was to be learned about past peoples from their bathrooms than from their battle-fields.

The exhibition 'feature', under the title *Clean and Decent*, ended a successful run and was dismantled. My friends began to tire of their malodorous jests about my occupation. I had recovered from the strain of being closeted at Olympia with visiting Royalty—I too was a delicate subject. But no sooner had the pots and pans been packed up, than *things closed in* again—a publisher suggested that *Clean and Decent* might make a good book . . . I hope it has.

I must apologise to the several journalists, amateur and professional, who saw the Olympia exhibit and its descriptive booklet, and had all—by coincidence—been just about to write this very book themselves. They asked only for the free loan of my notes, library lists, books, drawings and photographs. I hope that my refusals were gracious.

The book is meant as a little social history, popular rather than technical. Only the more enlightened schools of architecture and building, who give the humanities due weight in the syllabus, are likely to make it a set textbook. It is meant to entertain, even if scholarship does keep breaking through. The bibliography has been kept short, but serious students who read all the works there listed will find all but a few of the sources; general readers will not be distracted by pages peppered with references.

The story has not been carried through to the present day. It ends arbitrarily at about fifty years back, except for a few threads that for completeness have been allowed to over-run. Although there have been developments in our lifetime, these have been fully described in the technical and in the popular press. There would be little point in summarising the current textbooks, and it would be impossible to preserve an even scale. The missing pages would cover a very short period in relation to the whole. The technical achievements of today may not seem quite so important to the future historian who sees them in perspective.

Acknowledgements

INEVITABLY, much of this book is a compilation from other writers. Of those listed in the Bibliography, four are owed special acknowledgement. H. A. J. Lamb's historical survey from *The Architects' Journal* was the framework on which the book grew, and his references have led to many profitable sources. Most writers on the subject during the past twenty years have owed a like debt to Mr. Lamb, who ought himself to have expanded his much-quoted article into a book. Siegfried Giedion's *Mechanisation Takes Command*, Havard's *Dictionnaire de l'Ameublement* and Fuchs' *Illustrierte Sittengeschichte* have been equally rich mines of material.

Of lesser extent, but demanding acknowledgement, have been the borrowings from Geoffrey Ashe's jolly *Tale of the Tub*, Stevens Hellyer's vividly-written books on plumbing of 1877–91, and G. R. Scott's *Story of Baths and Bathing*. The material on monastic sanitation is largely drawn from Crossley's fine book *The English Abbey*, and many of the mediaeval documents quoted are either from Salzman's works listed, or from E. L. Sabine's paper. F. W. Robins' *Story of Water Supply* has been a clear and reliable source in that field. Several old advertisements are quoted from E. S. Turner's *Shocking History of Advertising* or from its predecessor Sampson's *History of Advertising* of 1874. Details of the Romano-British house come from Ward's *Romano-British Buildings and Earthworks*, most of the material on Brighton from Osbert Sitwell's and Margaret Barton's *Brighton*, and some odd items of cosmetic history from N. Williams' recent *Powder and Paint*. A little material has been taken, and any notable thefts admitted in the text, from M. C. Buer's social history, John Pudney's sketch of water-closet history in *The Smallest Room*, and last but not least Reginald Reynolds' fragrant classic *Cleanliness and Godliness*.

It is all too easy to re-quote unread material taken at second-hand, and often difficult to verify it at the source. Where the sound practice of verification has proved too difficult, the original work re-quoted has not been listed in the short Bibliography, but is usually referred to by the later author listed there. Detailed requests for permission

ACKNOWLEDGEMENTS

to borrow material have been many, but have not been made in
every case; to quote one writer who has himself been robbed here,
'it has been assumed that a main object of all these writers has been
the spreading of knowledge'.

Thanks to the time-honoured firms who readily lent priceless catalogues dating from about 1850 to 1910 for the Olympia exhibit, it
has been possible for the first time to illustrate that rich period as it
deserves. Godfrey Harwood has volunteered copies from the unpub-
lished correspondence of his relative George Jennings I. Mrs.
O'Hanlon has kindly translated passages from the Swedish, and
Mr. and Mrs. Harwood from the German. Thanks are due also to
Alan Adams, Donald Cook, T. S. Denham, W. H. Godfrey, J. T.
Hayes, Mrs. H. L. Jarman, C. M. Mitchell, G. B. L. Wilson, W. T.
Wren; to the Curator of Historical Relics of the British Transport
Commission, the staffs of the London Library and the R.I.B.A.
Library, and the Directors of the Leeds and Leicester City Museums,
the London Museum and the Science Museum.

The sources of the illustrations are listed and acknowledged on
page 268.

I

Man Becomes House-Trained

Different Concepts of the Bath — No Orderly Progress — Social History Reflected in the Bathwater — Early Man by the Waterside — Sanitary Taboos — Bathwater Not Portable — Neolithic Latrines in the Orkneys — Pottery an Old Craft — Sudden Emergence of a 'Modern' Bath — The Palace of Knossos — Its Drainage — Its Latrines — The Queen's Toilet — The Caravanserai — Ancient Cities — Egypt — Its Baths — The City of Ahkenaten — Egyptian Latrines — The House of Nekt — Ancient Greece — Its Private Baths — Its Public Baths

THE bath has had very different meanings, purposes and methods at different times in history. In Greece it was an adjunct to gymnastics, brief, cold and invigorating. In Rome and Islam it meant relaxation, bodily refreshment and resultant well-being; the basic method was the use of steam and water in a succession of varying temperatures. It was a social duty, carried out in company. The Greek or Roman bath was only incidentally a cleaning process. *Sanitas* meant health, not the removal of dirt. The communal bath of the Middle Ages, and the Turkish bath in its many revivals in Europe, had a like purpose. In the mediaeval monastery, the bath was strictly a routine cleaning, not to linger over, not to be enjoyed, and perhaps to be imposed—icy cold—as a penance. At times the bath has been used as a symbolic rite, its pleasures and its cleansing being both purely spiritual. In the eighteenth and early nineteenth centuries in Europe the bath normally implied medical treatment; the bather was 'the patient'. From about 1860 the bath revived as a private routine cleaning, but with something of the monastic penance in the preference for cold water. With the coming of liberal hot water, it became once more permissible to enjoy one's bath, and to its purpose of routine cleaning there is added today a dash of Roman relaxation.

These concepts overlap in time, and often mingle. There is no clear Wellsian outline of progress: a living Englishman has complained of his Oxford college that it denied him the everyday sanitary conveniences of Minoan Crete. The fifteenth-century gentleman used the bath, but his seventeenth-century descendant did not. The monk of 1350 enjoyed more orderly plumbing, and had sweeter habits, than the Londoner of 1850. The Polynesian 'savage' was cleaner than either.

Inventors in this field often upset the historian by coming on the stage years before any orderly textbook can give their cue. Baths being durable things, and plumbing rather inflexible, obsolete installations survive to belie the dating of their fashion. For all these reasons, the tale of the tub cannot be neatly parcelled into periods and tidily labelled with dates. As in any social history, Kings and Queens make poor milestones: the bathers of 1840 and of 1890 have

little in common but the adjective 'Victorian'. Centuries fit the facts
no better: the first and second halves of the century are quite different
in both eighteenth and nineteenth centuries, as they look likely to
be again in the twentieth.

The temptation is to arrange events in a logical order that they
never followed; to impose a path of development on some period of
unrelated experiments. This temptation will sometimes be resisted
here, though not to the point of incoherence. Some of the wavering
patterns of English social history will inevitably be mirrored, as it
were, in the bathwater—or found locked in the water-closet. 'To the
historian', says Siegfried Giedion, 'there are no banal things.'

Early man seems to have lived by the waterside: the oldest
Palaeolithic implements are found in the river gravels. Whether he
bathed or not, he had to have water, and because he had as yet no
means of bringing it to his dwelling, he made his dwelling by the
river bank. He is unlikely to have cared whether he was dirty and
smelly, or not, but he must have discovered, if only by accidentally
tumbling in, the refreshing effect of cold bathing. Thereafter, however gingerly, he might sometimes venture into the shallows. The
source of the water was itself the bath and wash-basin. After the
stage of merely *finding* a good spot for bathing, would follow the
obvious idea of removing a few toe-stubbing boulders and prickly
branches and *making* a bathing-place. At about the same time—that
is to say, within a few thousand years or so—he would have discovered the use of running water to carry away excrement, and
would have made a convenience of the brook or river. From these
conflicting uses of running water would soon arise the discovery
that the washing- and drinking-place should be upstream from the
'convenience'. Sanitary planning had thus begun.

In a few more thousand years he would begin to recognise that
the beneficent provisions of nature have a divine origin. He would
come to feel that elements such as water represented powers greater
than himself. Priests who developed this idea were wise enough to
understand something of the effects of pollution, and to make magic
or religion the basis for sound sanitary taboos. They would not waste
breath explaining these effects. It was simpler to make the water
supply sacred, and to give warning that any interference with it
could arouse divine wrath. Hence many watery ceremonials and
superstitions. The most primitive race still alive, the Australian
aborigines, who are still in the Stone Age, hold just such beliefs in
connection with their sacred water-holes. So do the Derbyshire
dalesmen today.

Once the settlement moved away from the riverside because of crowding, or in search of better hunting-grounds, and water having to be carried became more precious, drinking and cooking naturally came first, and bathing would be rare. Man in all periods has been willing to walk miles for a drink, but not for a bath. No peoples who have had to fetch and carry their water have ever favoured bathing in it. Even over short distances the weariness per gallon suffered in carrying it outweighs the refreshment per gallon gained from bathing. Until a civilisation is sufficiently advanced to have slaves—or better still, water-pipes—baths at home are just not worth while.

Under the dunes of Skara Brae in the Orkneys, Neolithic stone huts have been found, with crude drains leading from recesses in the walls that are supposed to have been latrines. There are beds, shelves and a sideboard, all of stone, but nothing resembling a bath. Cave paintings and ancient tumuli have yielded no hint of bathing or washing. Archaeologists have not dug up from datable strata any orderly series of baths evolving progressively from some rudimentary clay or stone tub. By the time civilisation is ready for its first real bath, the necessary technique already exists. Pottery is one of the oldest crafts, and even a modern fireclay bath is little more than a large dish. No special new technical skill is needed to make a man-sized earthenware dish, though its firing needs some care and its handling some effort. All that is really required is a suitable customer for the product. When this customer appears, in the person of a king who can command adequate labour and cares enough for the delights of home bathing, the bath emerges well-developed at the outset. It is as if the history of the locomotive should start suddenly with the 4-4-0 'Duke' class of 1895, not with Hero's steam-engine or even with 'Puffing Billy'.

The first known bath, then, is no tentative or rude effort. It startles because its form, dating from about 1700 B.C. or some 3600 years ago, is almost identical with the form of today. It is moreover set in an elegant bathroom with efficient plumbing. When its picture is put alongside that of a bath of 1891 A.D., few will say instantly which is which.

This earliest of all baths comes from Crete. That comparatively small island, left on one side today by all the main lines of cultural exchange, was the starting-point of European civilisation. Tradition told of a great palace built for King Minos at Knossos by his great craftsman Daedalus; that same who flew with ill-fated Icarus. When Sir Arthur Evans brought the palace to light, the legend fell far short of the wonders actually found. The account of his discoveries fills six large volumes. Minoan skill in hydraulic and sanitary engineering far surpasses that of the Chaldeans, the Egyptians or the Greeks.

4

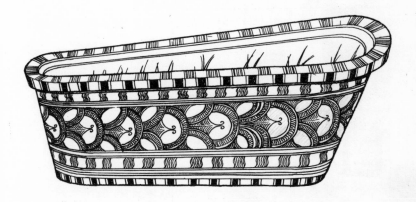

One of these baths is 3600 years older than the other. The clue as to which is which will be found on page 7

C.D.—B

Among the earliest elements of the Palace of Knossos is its water-supply through terra-cotta pipes. These are of tapering form, scientifically shaped to give the water a shooting motion that prevents sediment; in this respect Evans judges them better than modern parallel pipes. They are neatly interlocked and cemented at the joints. The 'handles' are not meaningless survivals from the similar vase forms, but were used to lash the pipes together to prevent displacement. These pipes date from about 2000 B.C. Each quarter of the palace had a drainage system of its own, passing into great main sewers, stone-built, large enough for the passage of a sewer-man, and flushed by the sometimes torrential rainwater. The Minoans evidently delighted in hydraulic devices, and used such refinements of the science as parabolic curves in the water channels and the precipitation of sediment in intermediate catch-pits.

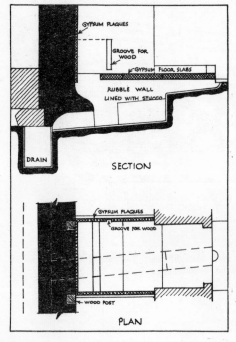

Latrine at Knossos

The Knossos latrines are remarkably 'modern'. One of these, on the ground floor, evidently had a wooden seat, and may have had an earthenware pan like a modern 'wash-out' closet, as well as a reservoir for flushing-water. Save for one short-lived water-closet of Elizabethan times, England had nothing comparable with this until the eighteenth century.

The Queen's apartments, with a private staircase leading to a secluded 'withdrawing room' and bedrooms above, a bathroom adjoining, and a short passage leading to a 'toilette chamber', had 'all modern conveniences' carefully planned and scientifically worked out. Two enclosed light-wells ensured fresh air and a diffused reflected light. The Queen's bathroom was lighted through an opening across which a translucent curtain may have been drawn. The painted terra-cotta bath has been restored and replaced in its probable position. Inside it are traces of a design with a suitably watery motif of reeds. It must have been filled and emptied by hand, but

The Queen's Bathroom at Knossos

A MINOAN INN

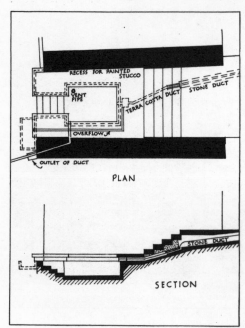

RECESS FOR PAINTED STUCCO

VENT PIPE

TERRA COTTA DUCT   STONE DUCT

OVERFLOW

OUTLET OF DUCT

PLAN

STONE DUCT

SECTION

*Foot Bath at the Inn
at Knossos*

8

there was a cistern in the 'toilette room' nearby, and a drained sink in the pavement into which the used bathwater probably went. A plaster dais or stand may have held ewers and washing basins. The walls are decorated with spiraliform bands, and in the 'toilette chamber' with a fine frieze of dolphins, in glowing colour; it is all fit for a queen.

Near the palace was a *caravanserai* or public inn. From its court-yard the tired traveller could go directly to a foot-washing bath, about 6 ft. by 4 ft. 6 in. by about 18 in. deep. The Minoans were a short race, and the average man could also have used it as a hip-bath. The surrounding slabs, jutting out over the bath, formed seats for the bathers. The water-supply, overflow and waste systems are most elaborate, involving no less than six separate ducts. The builder seems to have revelled in the use of pipes, but unlike the tangle some-times lovingly contrived by plumbers today, they were hidden. The bath, besides its inlet duct, has an overflow channel and a plug-hole with a stone plug. For private and more thorough bathing there were, in another room, terra-cotta bath tubs with decorative painting, with running water and a waste channel; there is some evidence that hot bathwater was provided. One of these bath tubs, about 4 ft. 8½ in. long, has carrying handles, and 'rowlocks' that evidently took a cross-bar to carry some object useful to the bather. The decorative motif outside resembles that inside the Queen's bath.

The main legend of the Palace of Knossos has been proved true, and though we cannot deduce from this that Daedalus was the archi-tect and sanitary engineer, we may well believe that if he was, his proved skill was such that he and Icarus may in fact have made that flight.

Bath in the Inn at Knossos

A square 'bath' found at Gaza, if correctly identified as such, and if correctly dated at 3100 B.C., is older than the Knossos baths, but this may have been only a storage tank. Palace bathrooms at Mari in Syria have been dated at about 2000 B.C., about the beginning of the Middle Minoan period when Knossos had its water-supply and probably had early baths now lost.

In the ancient cities of the Indus valley, flourishing from about 2500 to 1500 B.C., many houses had bathrooms and water-flushed latrines. The waste water went to the street drains by way of brick-lined pits, with outlets about three-quarters of the way up, rather like modern septic tanks and grit chambers. The earthenware pipes, latrines and masonry sewers of some Mesopotamian cities of about 1500 B.C. are still in working order. A neo-Babylonian 'bathroom' (which is itself the bath) has brick walls and a gypsum slab floor, both rendered waterproof with mastic (bitumen mixed with sand and fillers, trowelled on while hot). Its vertical drain is built of pottery rings.

Though all the peoples of the 'Fertile Crescent' (Assyrians, Babylonians, Sumerians, Hebrews and Syrians) used cosmetics and perfumes in quantities, they seem to have loved bathing less than the Egyptians, and their bath tubs were confined to the rich; only on festal occasions was the whole body washed.

Ancient Egypt has less exciting things to show than Knossos, but its best sanitary standards were fairly clean and decent. We gather that about 1491 B.C. Pharaoh took a daily bath in the Nile, for *Exodus* vii, 15, tells us that the Lord said unto Moses:

Get thee unto Pharaoh in the morning; lo, he goeth out unto the water; and thou shalt stand by the river's brink against he come.

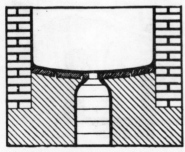

Neo-Babylonian Bath

A bas-relief in a tomb at Thebes shows an Egyptian lady at her toilet; the liquid that is being poured on her head seems too liberal to be scent or hair-dressing, but so meagre a ration can hardly be washing water, especially as the lady is wearing her head-dress, collar and jewellery. A small bathroom of about 1350 B.C. excavated at the City of Akhenaten at Tel-el-Amarna has a limestone slab across one end, with a low rim so that it holds only an inch or two of water. Low upright slabs form splash-backs. Water was presumably poured over the bather from a vase; it is really only a simple shower-bath. The waste water runs away through a trough in the wall and into a sunk vase that acts as a soakaway, or perhaps to save the precious liquid for the gardener. The contemporary latrine has a chilly limestone seat with a keyhole-shaped orifice; the soil falls into a removable vase in a pit below; perhaps another perquisite for the gardener. There is no water-flushing. In the 'House of Nekt' in the same city a different model is installed, with two brick supports that may have carried a wooden seat, Nekt perhaps believing like Dr. Johnson that 'the plain board is best'. In a third example, there are hollow spaces on either side of the seat that may have contained clean sand.

Limestone seat, Tel-el-Amarna

Homer often refers to the private baths of ancient Greece; in the *Odyssey* the weary traveller is nearly always greeted with a bath, often a hot one. The water is heated in a three-legged bowl set by the fire, with fresh wood burned beneath it. This is carried to the bath-house, and there 'diluted to a pleasant warmth' in a metal bath. Beautiful Polycaste, the youngest daughter of Nestor, gives Telemachus his bath,

washing him and anointing him with rich olive oil before she draped him in a seemly tunic and cloak: so that he came forth from the bath-cabinet with the body of an immortal.

At Menelaus' house, Telemachus and Peisistratus

went to the polished bath tubs and bathed; or rather, the house-maidens bathed them and rubbed them down with oil.

'Baths' seen in Greek vase-paintings are rather like large bird-baths and can hardly have been used to recline in. One picture shows a shower-bath fed by gargoyles on surrounding columns. When Agamemnon returned home from the siege of Troy and took a well-earned bath, his wife chose that opportunity to slay him by hitting him *twice* with an axe. Geoffrey Ashe acutely reasons that Agamemnon's bath can not have been a reclining tub, or she would have been able to do the job in one. Over-indulgence was frowned upon: Demosthenes grumbles about the habits of a lazy ship's crew who were 'washing all the time'.

The Greek public bath is only an adjunct to the gymnasium. It is in some ways like the Roman bath, but without its warm and hot rooms and waters or its leisurely enjoyment. The Greek public bath, whether shower or plunge, is brief and cold; for athletes rather than for sybarites. Hesiod condemns the use of hot water as unmanly. Only the later Greeks habitually yielded to this temptation, and then unlike the Romans they took the cold bath before the warm one. In decadent circles, games of draughts or snakes-and-ladders came to be preferred to athletic exercises at the baths.

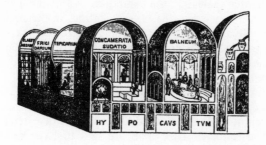

II

'Omnia Commoda'

WIDE-SPAN roofs of the past tell us one important thing about the people who built them. The purpose that they thought deserved a great enclosed space, and the technical effort of roofing it, is likely to have been their main interest in life. Today it is the aircraft hangar, with an even greater span than the exhibition hall or the cinema. In the nineteenth century it was the railway station; in the eighteenth century the noble mansion; in the Middle Ages the cathedral. In Rome it was the public bath. Here was the focus of communal life. Bathing was a basic social duty. The highest architectural and constructional skills were devoted to its setting.

The colossal scale of these baths is difficult to grasp. The Baths of Caracalla covered a site about 1100 ft. square, more than six times the site area of St. Paul's Cathedral, and could take 1600 bathers at a time. The Baths of Diocletian are said to have had twice this capacity: the vestibule alone sufficed Michael Angelo for conversion into the great church of *S. Maria degli Angeli.*

Rome was supplied with water by thirteen aqueducts of which the longest ran for about fourteen miles. Knowing tourists often point out that all this masonry was quite unnecessary, and that the silly Romans should have known that water finds its own level: a pipe run across the valley would have served. But the Romans were not so ignorant of hydraulic principles; they simply had no metal suitable to stand up to the pressure, such as bronze, in sufficient quantity to make such very large pipes. They well understood the relative costs of materials and labour.

The remaining overhead aqueducts have made such a dramatic impression on travellers, that it is not generally realised that the course of the Roman aqueducts was mainly underground. About A.D. 52 the total length of the eight main aqueducts was about 220 miles, of which only about 30 miles ran above ground.

In the fourth century A.D., Rome had 11 public baths, 1352 public fountains and cisterns, and 856 private baths. Some private houses at Pompeii had as many as 30 taps. As well as private water-flushed latrines, there were plenty of public ones; Rome in A.D. 315 had 144; in Puteoli there was one for every 45 persons, and in Timgad one for every 28.

At the peak Rome supplied 300 gallons per head per day. In London today we use about 51 gallons per head per day, of which 34 are for domestic and 17 for trade use; they must have wasted more than we do, but even so must have used more, especially for bathing.

14

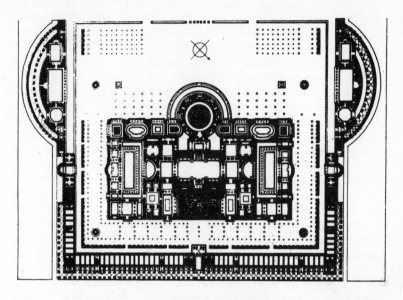

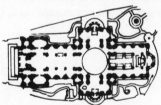

The Baths of Caracalla, Rome, and St. Paul's Cathedral, London, drawn to the same scale

The Roman bath drill, with some variations, is as follows. When the *aes* rings to announce that the water is hot, at about one o'clock, enter and pay your *quadrans* or quarter of an *as* (about half a farthing). Have a game of tennis in the *sphaeristerium* to get well warmed up. Enter the *tepidarium*, a moderately warm room, and sweat a little with your clothes on. Then undress in the *apodyterium* and get anointed. Remember Hippocrates' advice that

the person who takes the bath should be orderly and reserved in his manner, should do nothing for himself, but others should pour the water upon him and rub him.

15

You may bring your own special oil and ointment if you wish, but do not expect soap. The oil may be mixed with African sand if you are very dirty, as when you have been working on your chariot. Next move into the *calidarium* or hot room and sweat liberally, then briefly and more profusely in the *laconicum*, a hot spot directly over the *hypocaustum* or furnace, where the hot air is controlled by a sort of damper. Now have plenty of water poured over you from your head. There are three coppers for this water: warm, tepid and cold, to be used in that order. These coppers are connected so that as fast as warm water is drawn off it is replaced by tepid, and tepid by cold, a means of fuel-saving only lately introduced into modern furnaces. You are now thoroughly scraped with a *strigil*, a curved metal tool with a groove to collect the surprising amount of matter that will be removed. You are sponged. Re-anointed, you may end with a plunge in the cold bath in the *frigidarium*, before strolling or sitting around to meet your friends. You will feel splendidly refreshed and may well remark on the sorry state of the dirty native British in that newly-won island of woaded savages somewhere up north.

Opening hours at the baths were usually from 1 p.m. until dark, though one of the later Emperors had them lighted up at night. The bath was supposed to promote appetite, and some voluptuaries took one or more baths *after* dinner to enable them to begin eating again,

Roman Bath Implements: Strigils, Oil Bottle and Drinking Cup

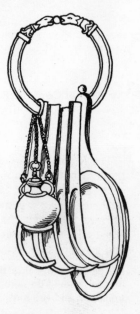

which occasionally proved fatal. Some of the more effeminate Emperors are said to have bathed seven or eight times a day. The last word in luxury were the *pensiles balneae* mentioned by Pliny, small baths suspended by ropes from the ceiling in which one could rock and roll.

In early times the sexes did not mix—even a father could not bathe with his son—and no respectable matron would go to the baths at all, but latterly promiscuous bathing was common.

On the walls of Pompeii are still various advertisements, including one that announces the opening of certain new baths:

DEDICATONE . THERMARUM .
MUNERIS . CNAEI . ALLEI .
NIGIDII . MAII . VENALIO .
ATHELAE . SPARSIONES . VELA .
ERUNT . MAIO . PRINCIPI .
COLONIAE . FELICITER .

To construe: there will be a dedication or formal opening of the baths, and the public are promised a slaughter of wild beasts, athletic games, awnings to keep off the sun, and perfumed sprinklings. These last (*sparsiones*) produced a *nimbus* or cloud of perfume to sweeten the spectators, all too necessary if they were such a rabblement as Casca tells of in Rome, who

hooted and clapped their chopped hands and threw up their sweaty nightcaps and uttered such a deal of stinking breath . . . I durst not laugh, for fear of opening my lips and receiving the bad air.

The method of spraying is supposed to have been some form of large syringe or other 'rude engine'. Another Pompeian poster reads:

THERMAE
M. CRASSI FRUGII
AQUA . MARINA . ET . BALN .
AQUA . DULCI . JANUARIUS . L .

offering warm, sea and fresh water baths. As provincial shopkeepers still like to have a London or Paris address, so the Latin provincials often added that they followed the customs of Rome. Thus the keeper of a bathing house at Bologna advertised:

IN . PRAEDIS .
C . LEGIANNI . VERI
BALNEUM . MORE . URBICO . LAVAT .
OMNIA . COMMODA . PRAESTANTUR .

his baths being in the fashion of the town, and offering every convenience. In the Street of the Fullers in Pompeii occurs the following, painted in red over another advertisement that has been whitewashed over:

IN PRAEDIS . JULIAE . S . P . F . FELICIS
LOCANTUR
BALNEUM . VENEREUM . ET .
NONGENTUM . PERGULAE
CENACULA . EX . IDIBUS . AUG .
PRIORIS . IN . IDUS . AUG .
SEXTAS . ANNOS . CONTINUOS .
QUINQUE .
S . Q . D . L . E . N . C .

Which has been translated:

On the estate of Julia Felix, daughter of Spurius Felix, are to let from the 1st to the 6th of the ides of August, on a lease of five years, a bath, a brothel, and 900 (90?) shops, bowers and upper apartments.

The seven final initials are said to mean 'they are not to let to any person practising an infamous profession', but this seems an odd clause where there is a *venereum* to let, and other erudites have seen in it *si quis dominam loci eius non cognoverit* and fancy that they can read underneath *adeat Suettum Verum*, translating as 'if anybody should not know the lady of the house, he should apply to Suettus Verus', all of which shows just how scholarly scholars can be. They have also translated the Pompeian street sign that takes the form of two sacred serpents: it means COMMIT NO NUISANCE.

Wherever the Romans settled they built public baths; there are remains of small Roman baths in London, and of an impressively large one at Wroxeter (*Uriconium*). Wherever they found hot springs

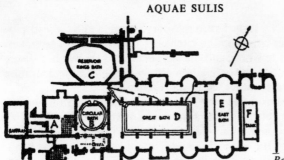

Roman Baths at Bath

they used them, as at the *Aquae Sulis* of Bath, where they began to build within ten years of the occupation (about A.D. 54). On the plan, A is a *calidarium* with sub-floor heating. The circular pool B, 33 ft. across, was perhaps reserved for women and children. C is the King's Bath, much used in the seventeenth and eighteenth centuries. The Great Bath D is 80 ft. long, 40 ft. wide and almost 6 ft. deep. The Roman bases still existing formerly carried columns about 40 ft. high, twice the height of the modern columns. These carried a vaulted roof of hollow tiles. The hot water still enters through a part of the lead conduit laid down by a Roman plumber nearly 2000 years ago. The Diving Stone still shows the impression worn by the feet of the Roman bathers. The two small baths E and F were not uncovered until 1923. The King's Spring rises at 120° F., and with two others gives a supply of half a million gallons a day. The temperature and quantity never vary. Such baths are but little echoes of the vast halls of Rome, but large or small, public or private, the principle is the same.

The term *villa* as applied to the Roman country house is inaccurate: the *villa* was the whole estate including the house. Most of the country houses excavated in Britain have been large—which is why their remains have survived—but the land was never studded with palatial mansions. Most houses were much smaller. In the towns few houses would have baths, because public baths were provided. The presence of a 'hypocaust'—a heating chamber under the floor, seen now as a series of rows of rough brick piers—need not argue a bath; this may have been only for sub-floor heating of an ordinary room. The remains are slight and the form of the houses found must be deduced from their foundations alone. They may have been very like some better-preserved examples found in Italy, though smaller and less sumptuous. In a Roman farmhouse at Boscoreale near Pompeii, that may give a fair idea of the British counterpart, the plumbing was found intact in every detail. The furnace that heated the hypocaust and the water tank was in a small room off the kitchen. The tank

19

was a lead cylinder, lagged with masonry. It was fed by a lead pipe from a cold-water cistern in the kitchen. The *alveus* or hot bath, surrounded by steps to sit on, and the *labrum* or cold shower bath, were so connected that by turning stop-cocks water of any temperature could be supplied to either, or to the kitchen.

This Pompeian example may help to elucidate the plan of a compact bath-house uncovered at Caerwent in Monmouthshire. This seems to have formed a semi-detached block at the west end of the house. The use of the rooms is somewhat conjectural but was probably as follows. Room A is an ante-room entered from an open court. On its left or south side is a cold-water bath B, 10 ft. 6 in. by 5 ft. 6 in., with a flagged floor bedded in concrete, 3 ft. below that of the room. Its sides are of fine brick concrete painted red. Between the bath and the room is a sill or dwarf-wall 9 in. high, with a step or seat in the bath. The plug-hole is in the middle of the south side. The second room C is a combined cooling- and dressing-room, with an alcove providing space for a lounge. The large room D is the warm room. The narrow room E is the hot room, and has a hot-water bath at the west end, smaller and shallower than the cold bath. The three walls flanking the bath are lined with hollow flue-tiles connected with the hypocaust below. The bottom of the bath is a single warm flagstone resting on the hypocaust piers. The last room F, the *sudatorium* or sweating-room, is immediately over the furnace, which is fired from a sunk yard or shed at G. There was probably a boiler here, as at Boscoreale; the supports remain. All the rooms have ¼-round skirtings of stucco, and these pass round the openings, suggesting that curtains were used and not doors.

At Lullingstone in Kent, a dig still in progress shows that the

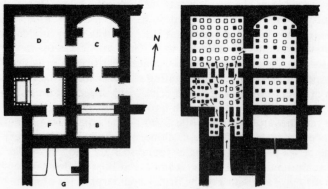

Bath House at Caerwent.
Ground Floor Plan   *Plan Through Hypocaust Below*

20

Roman colonist ran a self-contained community very like that of his later counterpart the country squire. The Archaeological Correspondent of *The Times* describes the findings up to 1957:

A plunge bath had been provided in the second century in the bathhouse almost beneath the ancestral elm tree—a noble giant craving also for water. In the fourth century this bath-house was repaired and enlarged. A sill exists still on which men sat and dangled their feet before plunging in. Another, smaller, by its side must have been meant for children.

On the opposite side of the site, and dug out of the hillside which had collapsed on top of it, the owner of this villa in the second century had added a household laundry. All the essentials of an efficient service were present. A rough mound at the back, built of brick, tile, and stone without mortar, was available either for bleaching the linen or hanging it out before washing. A tank, provided with hot water from a furnace close by, was used for treading out the dirty clothes, as is done in many a Continental town to-day and was done by Nausicaa and her maidens in the *Odyssey*. There was a hypocaust and additional heater or boiler. A drain behind carried off the water.

The holes still remain into which stakes were driven for a portable beehive clothes-airer.

. . . We are in the presence (we can feel) not of ghosts but of men and women with 'like appetites' to ourselves, whose dogs were friends, whose houses and their planning reflect good cheer and civilised habits, to whom a country life offered a satisfaction of its own—the more if, like Horace, they escaped to it from the sophistication of the town.

Not only the rich were catered for; the infantryman, marching every mile of the way from Italy and grousing at the cold mists of Northumberland, might find a hot bath ready when the ranks fell out. The fort of Housesteads, on Hadrian's Wall, had an elaborate system of stone water tanks, drains, a bath-house that has yet to be fully revealed, and latrines. A suite of bathrooms was provided for the barrack blocks. Surplus water from a tank flushed the latrines, and a tap allowed an occasional more thorough flush. The main latrine measures 31 ft. by 16 ft. internally. Along the two sides was a continuous trough, above which wooden seats were doubtless ranged.

Earthenware pipes (*tubuli*) were common. Those found in the kilns at Holt in Cheshire were tapered at the spigot end to fit into a large open socket. Wooden pipes were more rare. These sometimes had taper-and-socket joints; sometimes butt joints held by iron hoops, of lesser diameter than the pipes, with sharp edges so that they could be driven into the pipes, a mid-rib or stop ensuring equal penetration. Lead pipes were usually oval or ovoid, of sheet lead formed into a tube and soldered. Some of the pipes to be seen at Bath are oblong in section, the seams being joined by turning over the edges without soldering. Others are triangular and joined by a leaden rib. A length of about 10 ft. was normal. The *plumbarii* used lead from the Mendips, from Cumberland and from Wales. They often marked the product with their names, or with the names of the owners; pigs of lead have been found bearing the name of Hadrian. An important pipeline would have a tank every three or four miles so that repairs need not interrupt the supply, as well as to control the pressure, a practice resumed in the Middle Ages. Fittings such as stop-cocks were of bronze. Spouts usually took the form of animals' heads: dolphins seem to have been especially popular, and recur strangely in lands as far apart as Nepal, Tibet and Java. The Romans had pumps: one made on the lines described by Vitruvius in the first century A.D. was found at Silchester, of wood with lead cylinder-linings, though bronze was the usual material.

In Roman Britain there were no overhead aqueducts but there were 'leats' or open channels following the land contours. At Great Chesters on Hadrian's Wall a leat over 5 miles long brought water from Caw Burn, 2¼ miles away as the crow flies. Dorchester (*Durnovaria*) was similarly served. Lincoln (*Lindum Colonia*) had a conduit of earthenware pipes. At Birdoswald on Hadrian's Wall traces have been found of a charcoal filter-bed. Country villas might have storage tanks fed from leats, as at Chedworth in Gloucestershire where an architectural open basin held some 1500 gallons.

We must remember that the Roman occupation was not a single episode, but covered a period about equal to that between Queens Elizabeth I and II. It seems incredible that a civilisation so long established can have been so utterly effaced that its every art, custom and habit of living vanished, leaving no visible trace but the decaying roads. The explanation must be that the Saxons, Danes, Jutes and others who came to fill the vacuum did not come to rule and tax, but to destroy and supplant the natives. The Saxons seem to have disliked towns, and are supposed to have left London deserted. Even the earliest invaders do not seem to have occupied the Roman villas —perhaps they thought them haunted? They may not even have understood what the bath-house was for. Probably they razed the buildings to the ground, and then learned nothing even from the piles of bricks; if the simple art of brickmaking could be wholly lost, as it was, the art of plumbing could hardly survive. When the legions marched away, they could not take their baths with them, but they might as well have done so, for all the further use that was made of them. As Lord Grey might have said, the taps were being turned off all over Europe; they would not be turned on again for nearly a thousand years.

III

The Odour of Sanctity

Many writers repeat the tale that the early Church condemned bathing, quoting St. Benedict's command that 'to those that are well, and especially to the young, bathing shall seldom be permitted'. There is much evidence either way. St. Agnes died unwashed at the age of thirteen, and a fourth-century Christian pilgrim to Jerusalem boasted that she had not washed her face for eighteen years for fear of removing the holy chrism of baptism. The Blessed St. Jerome rebuked his followers for keeping too clean. But Gregory the Great, who was the first monk to become Pope and so would be used to the monastic sanitary routine, allowed Sunday bathing, and recommended baths so long as they did not become 'time-wasting luxury'. St. Boniface forbade mixed bathing in 745, and the public baths were indeed called *seminaria venenata*—hotbeds of vice—but it was to the sinning and not to the scrubbing that the Church objected. Pope Adrian I in the eighth century recommended the clergy to visit the baths in procession every Thursday. Archbishop Bruno of Cologne in the tenth century, and Adalbert Archbishop of Bremen in the eleventh, both abstained from bathing, but as an act of self-denial, as likewise a Frankish noble who, as penance for a crime, was forbidden to wash. For the other side: St. Francis of Assisi, although he spoke of 'our Sister Water, very serviceable and humble and precious and clean', listed dirtiness among the insignia of holiness, and St. Catherine of Siena not only avoided washing, but practised another and very costive form of self-denial.

Through those dirty days when 'for a thousand years Europe went unwashed', the monasteries were the guardians of culture—and of sanitation. They were the post-Roman pioneers of water-supply and drainage. A restive monk might be sent to a cold bath (which had some practical value in cooling the passions) and we know that Aldred, the chronicler of Fountains and Kirkstall Abbeys, when he had 'worldly thoughts', used to sit up to the neck in cold water, but a bath as such need not be a penance: the well-behaved monk would have a warm one.

24

The Monastery of St. Milburga, Much Wenlock, Shropshire: The Laver as existing in the twelfth century

For this a routine was laid down in the English monastery. The Chamberlain had to provide fresh straw for the mattresses once a year. He had to buy wood to keep up the fire in the 'calefactory' or warming-room, and to provide warm water and soap for baths two, three or even four times a year, for head-shaving every three weeks and for foot-washing on Saturdays. At head-shaving the monks sat silently in two rows in the cloister, facing each other; the older monks were treated first, and the youngest novice last, when the water was cool and the towels wet. The bath-house had plain round or oval wooden tubs of oak or walnut, and 'sweet hay' served for bath-mats. Extra baths might be prescribed for some of the sick brethren in the 'farmery'.

Near the 'frater' or refectory was the 'laver' or lavatory for a cold wash in a stone trough before and after each meal. This was in charge of the Fraterer, whose duty it was to see that it was kept clean. The laver was set back in the cloister wall clear of the walk, as at Norwich, or in a vaulted room alongside the cloister, as at Gloucester, or in a free-standing building in the 'garth' or garden court, like those at Canterbury and Durham, or the beautiful but sadly-ruined marble laver at Much Wenlock. A recess was provided for towels; the brethren were 'not to blow their noses on the towels or to remove dirt with them'.

25

Elaborate water-supply systems were made at surprisingly early
dates. At Christchurch Monastery at Canterbury a complete service
was laid on in 1150; the plans still exist. Close to the source was a
conduit-house—a round tower—from which an underground lead
pipe passed through five oblong settling-tanks to purify the water,
each with a 'suspirail' or vent to control the pressure; thence under
the city wall and into the monastic precinct. It then ran to a laver,
where it fed a tank raised on a central pillar to give a head of water.
From this ran two pipes: one to the frater, scullery and kitchen; the
other to the bakehouse, brewhouse and guest-hall, and to another
laver near the infirmary. In the lavers, thin streams of water trickled
constantly into the basins. Further branches fed the bath-house and
a tank for the use of the townsfolk. The waste ran into a stone fish-
pond, and from there to a tank by the Prior's chamber and so to
'the Prior's water-tub', where it was joined by the waste from the
bath-house and the rainwater from the roofs, to provide a hearty
cleansing flow through the main drain running under the 'rere-
dorter' latrines. There was even an emergency supply: in the infirmary court was a well, and beside it a hollow column standing on
the main pipe, into which well-water could be poured to keep up the
supply in times of drought. Short branches called 'purgatoria' served
for flushing the pipes. The efficiency of this water-system may
explain—as much as the holiness of the inmates—the exemption of
the monastery from the Black Death in 1349.

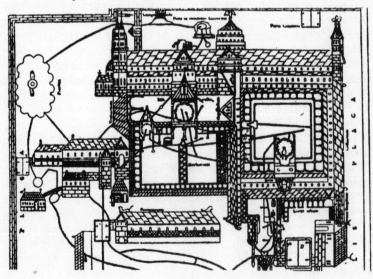

Plans of the water-supply and drainage system of the Cathedral and Monastery at Canterbury, completed c. 1167

Some of these early water-supplies meant deep digging, as shown by the tale of one of the miracles worked by St. Thomas à Becket. William of Gloucester, in charge of a scheme for Churchdown, was handling lead pipes in a cutting 24 ft. deep when the sides caved in. The local priest was celebrating Mass for his soul when St. Thomas intervened with a miracle, whereby William was dug out alive after a whole day's burial.

Some impressive distances were covered by the supply pipes: two miles at Bury St. Edmunds; more than three miles at Chester. Gloucester and Reading alike had two supplies: one for drinking and washing, and one for flushing the drains. At Reading the water was brought across the River Kennet by a 2-in. lead pipe, and at Lacock Abbey likewise under the Avon; a part of this pipeline was still in use in 1941. These pipes, like those of the Romans, were formed from sheet lead and soldered throughout their length, not drawn from solid. Sometimes, as at Beaulieu where the conduit-house still stands, hollowed elm trunks were used for the main supply; the museum there has earthenware pipes very like the Roman kind, as well as lead ones. The Sherborne Abbey conduit still supplies the swimming bath of Sherborne School.

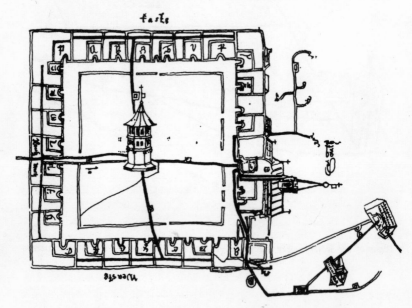

The 'Watercourse Parchment', London Charterhouse, c. 1430

The plumbing of the London Charterhouse is shown on the 'Watercourse Parchment' to be seen there in the Muniment Room. In 1430 one John Feriby and his wife Margery granted to the Prior and convent a 'fountain' (spring) and a portion of land across Islington to make an underground conduit. From the spring the water came by lead pipe and stone gutter, with 'suspirails' as at Canterbury. Crossing a ditch, the pipe was 'kevered with a creste of oke', and under a road it was 'closid in hard stone'. From a conduit-house in the Great Cloister, branches served the laver, laundry, buttery, brewhouse and 'Egipta the fleyshe kychyn'. Because the Carthusian monk lived almost isolated, doing his own cooking and keeping his own little garden, each cell had its own private water-supply. An extra pipe served two nearby taverns, the 'Windmill' and the 'White Hart', but this was only 'by sufferance of the Charter-house', as was proved in 1451 when the brewers brought a lawsuit which failed to establish their right to the water.

The only fixed monastic bath that is known, as apart from the mere tub, is a remarkable one found at Kirkstall Abbey in excava-tions made since 1950. This is a stone-built chamber sunk in the ground in an open courtyard. It is close to the angle of the courtyard formed by the warming-house and the dorter sub-vault. It measures about 4 ft. 7 in. by 4 ft. 3 in. and must originally have been at least 5 ft. 6 in. deep. The walls and floor are of roughly-faced sandstone, and steps descend into one end. A lead pipe supplies it and it has a stone drain; a removable stone serves as a bath-plug. The evidence dating the bath to the thirteenth century seems conclusive. It has now been extensively rebuilt with new stonework, and looked very different when first uncovered. If the attribution is correct, this bath is unique in Great Britain and perhaps in Western Europe. There remain some odd things about it that may still justify some uncer-tainty. Its placing is strange: it is less than a pace away from the sub-dorter doorway (of about the same date) and almost as wide as the doorway, though not quite central with it. It would be difficult to emerge from the sub-dorter in a hurry without falling into the bath unless it were covered; indeed, in this it is reminiscent of the pool facing the hotel doorway in Chaplin's old film *The Cure*. Its shape and depth seem strange for a bath. The stones are not set in mortar and it would not hold water for very long. It was at first thought to be a cold-storage chamber for meat, milk or wine, but this was before the plumbing was discovered. Until further evidence appears, the verdict on the Kirkstall bath may perhaps be 'not proven'.

An account of the monastery at Durham written in 1662 describes both laver and privies:

Within the Cloister-garth, over the Frater-House-door, was a fair Laver, or conduit, for the Monks to wash their hands and faces, being round, cover'd with Lead, and all of Marble, saving the outermost walls, within which they might walk round about the Laver. It had many Spouts of Brass, with twenty-four Brass Cocks, round about it, having in it seven fair Windows, of stone work, and over it a Dove-cote, cover'd with Lead, finely wrought; as appears to this day.

Adjoining to the East-side of the Conduit-door hung a Bell to call the Monks, at eleven of the Clock, to come and wash, before dinner, having their Closets or Ambries on either side of the Frater-House door, on the outside within the Cloister, kept alwayes with clean Towels to dry their hands.

There was also a large, and decent place, adjoining to the West-side of the said Dorter, towards the water, for the Monks and Novices to resort to, called the Privies, two great Pillars of Stone bearing up the whole floor thereof. Every Seat and Partition was of Wainscott, close on either side, so that they could not see one another when they were in that place. There were as many Seats on either side as there were little Windows in the Wall to give light to the said Seats; which afterwards were walled up to make the House more close. At the West-end of it there were three fair glass Windows; which great Windows gave light to the whole House.

Water Tower at Canterbury

Kirkstall Abbey

In many monasteries the 'rere-dorter' or sanitary wing, on an upper floor, was linked to the 'dorter' or dormitory by a bridge, serving—and perhaps intended—as the equivalent of the 'ventilated lobby' of modern Byelaws. These wings were often very long— 145 ft. at Canterbury, and 158 ft. at Lewes. Since every act of the day was subject to a strict time-table, large numbers had to be expected. At Furness the seats were ranged back-to-back in a double row, but usually they were set against the wall in one long single row, divided by partitions and each with its window, as at Durham (described above) and at Fountains Abbey in Yorkshire. Below was a walled-in drain that was either artificial, or a natural stream perhaps diverted. A stream for drainage was an important factor in siting a monastery. Where the conventual buildings are on the North side of the church, an unusual plan, a stream will often be found on that side. The stream was split up if necessary, so as to flow under the rere-dorter, infirmary and kitchen, and might later be covered in for some distance; many so-called 'secret passages', which sometimes —to the delight of heretics—seem to link monastery and nunnery, but tend to lead down to watercourses, are in fact old sewers of this kind.

Water-flushing from above was rare, but at St. Albans the Abbot built a stone cistern to store rainwater for this purpose, serving a Necessary House 'than which none can be found more beautiful or more sumptuous'. His predecessor had made, at Redburn Priory, his own private latrine, 'because formerly one building had served him and the brethren there, wherefore they were ashamed (*erubescebant*) when they had to go to the Necessary in his presence'. Of excavations at St. Albans in 1924 A. J. Lamb writes:

Here was found a deep pit, 18 ft. 8 in. long by 5 ft. 3 in. wide; the walls were 15 in. thick. The depth of the pit below the Cloister floor level was 25 ft. At the bottom were found pieces of pottery and fragments of coarse cloth which, it is thought, were old gowns torn up by the monks and used as toilet paper. Evidence, too, that the monks suffered from digestive troubles, which were by no means rare in those days, was proved by the finding in the pit, seeds of the buckthorn—a powerful aperient.

At Tintern on the Severn the tide, says Reginald Reynolds, was 'ingeniously employed to effect a complete flushing of the monastic dykes that must surely have swept them from their seats when the tide was highest'.

The monks, then, had all the essentials of modern sanitation save running hot water, and may be rated clean and decent.

IV

'Every Satyrday Nyght'

*The Norman Castle — The Well — Washing Ritual at Table —
The Ewer — The Boke of Curtasye — Private Head Washing —
The King's Wardrobe — Mediaeval Wash-stands — Bath Tubs
— The Communal Tub — Bath, Food and Music — Royal Bath-
rooms — Pots for the Stews — Chaucer's Metal Bath — Coy
Evasions — Castle Garderobes — Moats and Pits — Cosy Re-
treats — Private and Public Latrines — Polluted Watercourses —
Sanitary Ordinances — The Gongfermors — Leonardo's Latrines*

IN the Norman stronghold, built to withstand prolonged sieges, a well was essential, not merely within the defences but inside the final refuge, the inner 'keep'. Even there, the stone lining of the well might be carried up to the first floor or higher, as an added protection against besiegers who might break in below. Water could be drawn at any floor level. At Newcastle, the well is in an angle turret of the keep, and there are basins or recesses in the walls on either side of the well-head, with pipes and channels running to other parts of the keep. From the end of the thirteenth century keeps were no longer built, and the well was in the central open space, or in a special tower as at Carnarvon, where the water was distributed from a lead-lined tank through stone channels.

In castle and manor-house, although the toilet routine was less strict than in the monastery, habits were more cleanly than has generally been supposed. Fingers came before forks, and every diner dipped into the common dish before sharing a helping with a table-companion; a slice of bread (*tranchoir* or trencher) served the purpose of a plate. This made hand-washing before and after meals almost unavoidable, a custom perhaps better observed then than it is today.

> Thai set trestes, and bordes on layd;
> Thai spred clathes, and salt on set,
> And made redy unto the mete;
> Thai set forth water and towelle.

Donner à laver or offering a wash became a polite ritual, strictly codified; *laver* meant only this formal washing at table, not the ordinary washing with soap in private (*se décrasser*). At the top of the table an important person might wash separately, but it was polite to share the bowl with one's neighbour, and pleasant to play

manus manum lavat (one's hand washing another's) with one of the
opposite sex. Neglect or refusal could be a calculated insult. Water
was poured over the hands from a jug while they were held over the
bowl. The water might be scented or strewn with rose-petals, but
soap (first manufactured in England in the fourteenth century) was
not brought to the table. Ornamental gold, silver or silver-gilt wash-
basins and jugs appear in early inventories; in 1365 the Duke of
Anjou had no less than sixty sets. The basin often had a raised boss
in the centre, enamelled with the arms of the owner, on which the
base of the jug fitted. Lesser persons used brass or pewter (an alloy
of lead and tin). A gentleman at Northallerton who made his will in
1444 had in his hall, among other utensils, four such basins and two
jugs for washing at table. Froissart, writing of the English court in
1350, tells of the King sharing his wash-bowl:

*Quand le souper fut appareillé, le roi lava et fit laver tous ses chevaliers,
si s'assit à table et les fit seoir de lez luy moult honorablement.*

35

A passage in the *Heptameron* shows that this courtesy need not be extended always to a lady:

Après que le gentil homme eut lavé avec le Seigneur de Burnaige, l'on porta l'eau à ceste dame qui lava et s'alla seoir au bout de la table.

The procedure may well have been going on unchanged for thousands of years; it is exactly the same as described in the *Odyssey*, in which Menelaus calls out 'Ho there, pour water once more upon our hands that we may eat!' and more than once Homer tells how before a meal 'a maid came with a precious golden ewer and poured water for them above its silver basin, rinsing their hands'.

An alternative to the mediaeval jug and basin was the double basin. The proper form for tendering this to the guest is fully laid down. The butler sets the basins on the sideboard, one being inverted to form a cover. The signal for washing is given by blowing trumpets or by music from the minstrels: in the metrical romance of *Richard Cœur de Lion* 'at noon à laver the waytes blewe', 'noon' being the canonical hour of 'none' or 3 p.m. The 'ewer' (a word for the servant, but sometimes applied as nowadays to the water jug) takes up the basins; the butler raises the cover and checks that the water has not been forgotten and is duly scented. The ewer kneels before the guest —a tricky operation for a small page-boy with a large basin—and takes the full basin in his right hand, the empty one in his left. The guest holds his hands over the empty one, and water is poured over them. The full basin may have a spout for this purpose. The towel is over the ewer's left arm. When all is done, the basins are handed to the butler—it is bad form to close them. If the guest is a very important person, the host may carry out the duties of the ewer.

In the 'Boke of Curtasye' other table rules are laid down: you should not play with cats or dogs at meal times; your nails should be clean:

> Loke thy naylys ben clene in blythe,
> Lest thy felaghe lothe therwyth.

You must not spit on the table:

> If thou spit on the borde or elles opone,
> Thou shalle be holden an uncurtayse mon,

and if you blow your nose (having at this date no handkerchief) you should wipe your hand afterwards on your skirt or your tippet:

> Yf thy nose thou clense, as may befalle,
> Loke thy honde thou clense withalle,
> Prively with skyrt do hit away,
> Or ellis thurgh thi tepet that is so gay.

36

Nor may you pick your teeth with a knife, nor clean them with the table-cloth. Another book of etiquette says that when you wash your mouth at table you must not eject the water back into the basin, but decently on to the floor.

The private 'head-washing' (*laver la teste*) was a more thorough operation than the term suggests. The utensils were much simpler than those used at table. The basin was shallow and quite large, usually of copper, brass or tin, sometimes of silver, highly polished inside so that it may have served also as a mirror. It was put on the floor, on a mat; stripped to the waist, the user knelt, and could take what a modern nursery has called *harfabarf*, or to complete the task he could sit in it. Such basins, with their jugs, are found in ancient Egyptian and Greek pictures. An ingenious example shown in an eleventh-century manuscript, probably of terra-cotta, has a long handle made hollow to serve as an emptying spout.

Such were probably the utensils provided for the Wardrobe at Westminster which is described, in an order of 1256 for pictorial decorations, as 'where the King is wont to wash his head', and for Edward IV, in whose Household Book it is laid down that

this barbour shall have, every Satyrday nyght, if it please the Kinge to cleanse his head, legges or feet, and for his shaving, 2 loves, 1 picher wine

—in which the conditional clause is noteworthy. But for private

Detail from Dürer, 1509

washing there were also fixed stone basins (*lavatoria, lavabos* or lavers) long before the coming of running water. These were sometimes called *bénitiers*, which are strictly holy water stoups, but were not necessarily for sacred use. Dürer, illustrating *The Life of the Virgin* in 1509, shows such a fixture in a niche, with a hook over the basin on which is hung a portable spherical reservoir with a carrying-handle and a tap. This could no doubt have been heated over a fire. He shows a long towel on a roller, exactly as used today except that the invention of the endless roller-towel seems just to have been missed. Dürer's study, preserved with furniture that may or may not be authentic, in the *Dürerhaus* at Nuremberg, has an almost identical laver and towel-roller, but it is doubtful whether these fittings have survived from 1509 and that the woodcut was drawn from them—they may be restorations based on the woodcut. Another mediaeval

38

form is a portable wash-stand of wood or metal, with a metal bowl, a reservoir and a tap, a shelf below for soap or toilet-water, and a towel-rail. A fourteenth-century illumination reproduced in Wright's *History of Domestic Manners* (1862) shows what the author describes as a nun 'arranging her lamp' before going to bed, adding that the use of the 'stand' under it is not easily explained. This is in fact a laver, and the 'lamp' is the reservoir.

These lavers could be decorative: at Westminster in 1288 payment was made for '5 heads of copper, gilded, for the laver of the small hall', and for 'the images of the said laver, and for whitening (tinning?) the laver and gilding the hoops'.

Mediaeval books of etiquette insist upon the washing of hands, face and teeth every morning, but not upon bathing, though when a visitor arrived it was good form to offer him a bath. An early writer objects to the foppishness of the Danes, who

following the custom of their country, used to comb their hair every day, bathed every Saturday, often changed their clothes, and used many other such frivolous means of setting off the beauty of their persons.

King John took a bath about once every three weeks, and his subjects presumably less often. The bath tub, like that of the monastery, was of wood; it is often found among the devices used as trade signs by the

The Mediaeval Wash Basin, c. 1500

coopers. Such tubs, usually round, were sometimes lengthened into a form not unlike the modern bath, not to let the bather lie down, but to make room for others. The communal tub had one naughty but one good reason; the good reason was the physical difficulty of providing hot water. No modern householder who, with a frozen bath-waste, has baled out and carried away some 30 gallons of water, weighing 300 lb., will underrate the labour involved. The whole family and their guests would bathe together while the water was hot. Many pictures of the communal tub show a tray across the bath, with a meal on it, and perhaps musicians to add to the fun. Many physicians protested against excessive drinking in the bath. Nobody was shy: a charming illustration to a thirteenth-century manuscript shows a knight who dismounts from his steed to attend his lady's open-air tub; another shows a knight in his bath served by young women who shower him with rose-petals, the mediaeval equivalent of bath-salts. In mediaeval stories of amorous intrigue, the two lovers usually begin their evening by bathing together. Ideas of propriety were different from ours; the whole household and the guests shared the one and only sleeping apartment, and wore no night-clothes until the sixteenth century. It was not necessarily rude to be nude. Home life seems to have combined luxury with discomfort, and a strange indifference to privacy.

The wooden tub often had a decorative and protective fabric canopy, and might even be padded inside with submerged linen. In 1403 Princess Marguerite of Flanders buys 64 ells of common cloth (*toille bourgeoise*) to pad two bath tubs, and red Malines cloth for a canopy. To fill or empty the tub, a small wooden baler was made exactly like a miniature tub, but with only one handle, formed (like those of the tub) by lengthening one stave. One illustration shows a wooden tub with a waste-pipe that must have saved much baling.

Salzman's recently published book *Building in England down to 1540* releases a wealth of fresh documents throwing new light on early baths, and his researches show that the hand-filled wooden tub was by no means the best thing available. Some royal bathrooms, at least, had fixed baths properly cased in, tiled floors, bath mats, heating systems and even running hot and cold water. At Westminster Palace in 1275 Robert the Goldsmith was paid 14*s.* for 'a key (tap) to the laver in the bath room' and 26*s.* 8*d.* for '4 keys of gilt bronze and 4 leopards' heads for the baynes (baths)'. By 1351 the Palace had 'h. & c.', when Robert the Foundour was paid 56*s.* 8*d.* for '2 large bronze keys for the King's bath-tub for carrying hot and cold water into the baths'. At Langley a 'square lead (tank) for heating water for the stues' is mentioned, and this water may have been piped directly to the bath. More often the water was carried in pots that may have been filled from the 'leads' or perhaps heated directly over the furnace. Thus at Bardfield Park, 41 earthen pots were bought for the *stues*, costing 18*d.* each, and at Windsor 125 pots at 8*d.* The furnace of the King's *stywes* at Windsor was remade in 1391, and John Brown the carter carried no less than 229 pots to the castle from Farnborough. At Eltham John Jury the Potter supplied 120 *pottes pro styuez*, and Thomas Mason was paid £4 'for making the walls and setting 2 leads called *fournaysez* and 120 pottes for the *stuuy* house'. In a Westminster account of 1325:

William de Wynchelse for 3 boards called righok for crests and filetts of the bathing tub,—18*d.* For 3 oak boards called clouenbord for making the covering (canopy) of the said tub, 6 ft. long and 2½ ft. broad,—3*s.*
. . . for 100 fagett (faggots) for heating and drying the *stuwes*,—3*s.* For a small barrell, 2 bokettes and a bowl for carrying water to the *stuwes* . . . carpenters working on the covering of the bathing tub and the partition in front of the said tub—for 6 pieces of Reigate stone for making a slabbing in front of the partition of the said tub in the King's ground-floor chamber . . . for 2250 pavingtil (paving tiles) for the said chamber . . . for 24 mattis (mats), at 2*d.* each, to put on the flore and pavement of the King's chamber on account of the cold.

44

45

The *stewes* or *stues* or *stywes* or *styuez* or *stuuys* were not the only item allowing some latitude in spelling: about the end of the fifteenth century pumps were coming into use, and have to be recognised as such in *the soker of the pimp, xlvj foot of pymps, a olde Water Pompe,* and even *a plump-maker.*

Some lines from Chaucer have been quoted as evidence that he knew of metal baths. In *The Second Nun's Tale* he tells of a bath in which 'St. Cecile' was

> ... fastè shetten,
> And nyght and day greet fyre they under betten.
> The longè nyght, and eek a day also,
> For al the fyre, and eeek the bathès heete,
> She sat al coold and felt of it no wo;
> It made hire nat a drope for to swete.

It is argued that a bath heated thus over a fire must have been of metal. But this St. Cecilia (and there were at least two) was a Roman lady of about A.D. 200, and the bath, as evidenced by the words 'fastè shetten' (shut fast) and 'swete' (sweat), must have been an ordinary Roman *laconicum* or hot room. Chaucer's metal bath will not hold water.

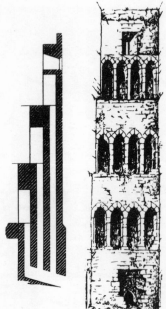

Garderobes at Langley Castle

46

In evading direct mention of the 'closet' the genteel English have devised many euphemisms. But it is a surprise to find this taboo already in force in mediaeval society; one would have supposed an earthier way of speech to have ruled, even among nobles and monks, than such coy evasions as 'necessarium' or 'necessary house' or, more oddly, 'garderobe' (wardrobe), exactly as we whisper today of the 'cloak room'. 'Garderobe' must be used here, being the accepted term for the mediaeval privy closet.

In the great house or castle the garderobes were usually built within the ample thickness of the walls, each with its own vertical shaft below a stone or wooden seat. Sometimes a lone garderobe would be formed within a buttress, as at Stokesay, or corbelled out in a projecting turret with an open drop instead of a shaft. A block of several might be repeated at each floor level, with multiple shafts grouped neatly like chimney flues. At Langley Castle in Northumberland, four seats on each of three floors were so arranged, and were thought worthy of the principal tower. At Bodiam Castle there were at least twenty seats to choose from. At Southwell Palace they radiated round a central shaft, facing outward on to a circular passage: neighbours were sociably within hearing but decently out of sight. As well as such groups, many castles had one for each

*Garderobes at
Southwell Palace*

47

important room, including one conveniently near the banqueting table, as in the Tower of London. This is a small vaulted chamber within the wall, about 3 ft. wide, with a narrow window. A stone riser across its width probably carried a wooden seat, and a short shaft turns outward through a hole to discharge down the face of the wall into the moat below. Formerly defensive, moats must have become offensive. In 1313 Sir William de Norwico ordered a stone wall to screen the garderobe outlets of the Tower keep.

Many of the 'hiding holes', 'priest's holes', 'oratories' and 'private chapels' beloved of National Trust guides were in fact garderobes, as for example at Abingdon Pigotts near Royston where the 'altar slab' in the 'chapel' has a hole in it.

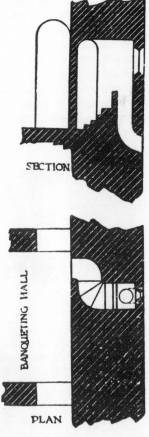

Garderobe in the Tower of London

Where no stream or moat ran below, there might be a removable barrel, or a pit like the great pit at Everswell ordered in 1239 for the Royal Privies, or that one dug at York Castle in the reign of Edward III at a cost of 10s. What the cleaning of these pits meant is shown by an account of the work at Newgate Jail in 1281, when thirteen men took five nights to clear the 'cloaca', at a total cost of £4 7s. 8d. This was then thought a high figure, but the men were paid three times the normal rate, pocketing 6d. each per night. Meanwhile four watchmen guarded the gap that had to be made in the wall, lest prisoners should think liberty worth such a noisome exit. Conversely, when Henry III ordered the Sheriff of Surrey to cause a privy chamber to be made adjoining the King's Great Chamber on the ground floor at Guildford, the Clerk of Works was to block up the outlet in the ditch with strong bars to deter intruders. King Henry was not above dictating to the Constable of the Tower such homely details as that he should 'cause the drain of our privy chamber to be made in the fashion of a hollow column, as John of Ely shall more fully tell thee'.

Such garderobes were often draughty (though as there were no flushing tanks this was perhaps as well), but they were sometimes built into chimney breasts for warmth, as was one for the King at Winchester. The frequent placing of the garderobe pit or drain beside the kitchen flue was probably due to this plan and not, as has been suggested, so that it might also receive kitchen rubbish. It must not be thought that the rooms in a mediaeval castle were always as bare, cold and bleak as they look now with only the structural stonework remaining. Any room in a modern flat that survives will look uninviting after centuries of desertion, its windows, plaster, woodwork and furniture gone and its concrete shell alone remaining. We can only guess how these garderobes were furnished, but a well-appointed one, wainscotted, matted, and perhaps with 'paper wallys' and a bookshelf, could have been almost cosy. In *The Life of St. Gregory* this is the retreat recommended for uninterrupted reading.

From the Close Rolls comes an order of 1246 from the King to Edward FitzOtho:

Since the privy chamber in our wardrobe at London is situated in an undue and improper place, wherefore it smells badly, we command you on the faith and love by which you are bounden unto us, that you in no wise omit to cause another privy chamber to be made in the same wardrobe in such more fitting and proper place as you may select there, even though it should cost a hundred pounds.

A palatial privy this, if the price—about £4,500 today—is to be taken literally.

Old London Bridge

Among the lower orders there were some private latrines, as we know from housing ordinances of 1189, requiring that garderobe pits, if not walled, must be at least $5\frac{1}{2}$ ft. from the party line; if walled, $2\frac{1}{2}$ ft. That there were public ones we know from the inelegant fate of John de Abydon, who was set upon by rogues and killed, in 1291, as he was coming out of a common privy in London Wall within Cripplegate Ward, at the head of Philips Lane. Others are recorded: a 'four-holer' at Temple Pier south of Fleet Street, built over the Thames and roofed in; another at Queenhithe over an open sewer; another serving the occupants of the houses on London Bridge. Of these houses there were 138, making for a busy privy, though many householders no doubt took advantage of the site to make simple private arrangements, which would add to the known risks of 'shooting the bridge', built as they said 'for wise men to go over and for fools to go under'. The London Bridge privy had two

50

doors, whereby it is recorded that a debtor escaped his attendant creditor one day in 1306.

London's watercourses, such as the Fleet River and the Walbrook, with latrines built over them, had long become offensive. The monks of White Friars protested to King and Parliament that the putrid exhalations of the Fleet overcame all the frankincense burnt at their altars, and caused the deaths of many brethren. Stow, who tells this, adds that in 1300 Sherborne Lane, once 'by a long bourne of sweet water', had become known as Shiteburn Lane. In 1321 complaints were made that a London lane called Ebbegate was 'blocked' by the effect of its overhanging latrines, *'quarum putredo cadit super capita hominum transeuntium'*. With such risks to be faced, we can like A. J. Lamb sympathise with

the ingenious woman who constructed her own privy and connected the outflow by a wooden pipe to the rainwater gutter. Unfortunately, the complaints of the neighbours, when the pipe became blocked, brought a heavy summons on the unlucky woman's head.

Fifteenth-century Latrine

51

About the mid-century, as the Black Death reached its climax and carried off a third of England's population, some attention was given to its causes, and the clearing of the 'sore decayed' Fleet was listed among the sanitary ordinances of Edward III, sixty years after the protest from White Friars. No more privies were to be built over the Fleet, and others only where they abutted on water-courses. Each owner was to pay 2*s*. a year towards the cost of their public cleaning. Cesspits were favoured; these were at least preferable to the Ebbe-gate system. In a contract dated 1370 for the building of 18 shops in London, the mason was to make '10 stone pits for *prevez*, of which pits 8 shall be double (i.e. serve two houses,) and each in depth 10 ft., and in length 10 ft., and in breadth 11 ft.'

Even these generously-dimensioned pits had sooner or later to be cleared by the *gongfermors*, a deservedly well-paid trade. At Queen-borough Castle, being rebuilt on up-to-date sanitary principles in 1375, this job was done by the happily-named partners William Mokkyng and Nicholas Richeandgood. The *gongfermors* seem to have become adapted to their horrid task, for an ancient story tells of one who complained, out of working hours, about the smell of a badly-snuffed candle. An account for some house repairs in 1450 includes for 'the dygyng of a pyt and takeyng owte of a serteyne of dounge owte of a privey and for to bery ye dounge in ye same pyt (5*s*. 6*d*.) . . . ye caryyng a way of vj ton of dounge (17*s*.)'

The same account includes for 'makynge a vente for a prevey to voyd ye heyr (air)', and it was becoming recognised that 'a wyse buylder wyll set a sege house out of the wey from syght and smellynge'.

Washing and Scourging

As early as 1449 one Thomas Brightfield of the parish of St. Martin built some kind of water closet, flushed by pipe water from a cistern, but this was much before its time and was not imitated. In 1579 Tower Street, All Hallows, still had but three privies for nearly 60 houses.

We should not judge the sanitary conditions of these times solely from their more unsavoury incidents. Public nuisances, complaints and lawsuits tend to fill the record, but for every offensive privy so immortalised, there may well have been ninety-nine decent ones.

That prolific inventor Leonardo da Vinci (1452–1519) in his proposal for Ten New Towns (a pre-echo of Trystan Edwards here) aimed to 'distribute the masses of humanity, who live crowded together like herds of goats, filling the air with stench and spreading the seeds of plague and death'. In these towns the drainage of all private and public privies, and all garbage and street sweepings, were to be carried to the river by sewers (*vie sotterane*). All the stairways in the tenement buildings were to be spiral, to prevent the insanitary misuse of stair landings. Leonardo invented a folding closet seat that 'must turn round like the little window in monasteries, being brought back to its original position by a counterweight'. He was equally inventive in the bathroom: when Isabella d'Aragona asked for one in the Corte Vecchia, he sketched a hot water system and provided for pre-mixed bathwater, calculating that 3 parts of hot water to 4 of cold would give the right temperature. For Francis I at Amboise Castle he proposed to instal a number of water-closets with flushing channels inside the walls, and ventilating shafts reaching up to the roof; and as people are apt to leave doors open, counterweights were to be fitted to close them automatically. This, like his projects for flying machines, parachutes, military tanks, machine guns and submarines, though of likelier benefit to mankind, remained no more than a project. Man had still a long time to wait for the water-closet.

V

Bath Knights and Bagnios, Conduits and Quills

The Ritual of Ablution — Pilate's Symbolic Act — Kali Catches Nala — Hindu and Brahmin — The Holy Ganges — Duckings in Lapland and Russia — Urine, Blood, Mud and Pitch — Contagious Holiness — Fishermen's Luck — Rain Making — Knights of the Bath — The Stews — Bagnio and Bordello — Decay of the Stews — Mediaeval London's Water Supplies — Wooden Pipes — Conduits — Water Rates — Bursts — Plumbers at Westminster — Dublin — Hull — Naughty William Campion — Conduits Ran With Wine — Waterworks at London Bridge — Bulmer's Engine — Myddleton's New River — Quills — The Worshipful Company of Plumbers

BATHING and washing have always had special ritual uses in religion, chivalry and magic, that have nothing to do with physical cleanliness. *Ablution* is the proper technical term. The meaning of the word is clouded by its modern misuse in camp and barracks: 'the personnel require proper ablutions for their personal hygiene', said a Flight Sergeant to the author, meaning that the men needed a place to wash. The aim of ablution is to remove, not dirt, but the invisible stains contracted by touching the dead, by contact with childbirth, murder, persons of inferior caste, madness or disease.

When Pilate saw that he could prevail nothing, but that rather a tumult was made, he took water, and washed his hands before the multitude, saying, I am innocent of the blood of this just person.

Had Pilate merely spoken the words, we might not have been able to absolve him as we do now, for want of the symbolic and memorable act that is the whole point of ritual.

According to Geoffrey Ashe the Brahmins tell a cautionary tale of King Nala of the Nishadhas: the god Kali, having watched him for twelve years, at last caught him neglecting to wash his feet at the proper time. Immediately he was a doomed man. Kali disguised himself as a set of dice, and next time Nala gambled, he lost everything but his loin-cloth. For thousands of years orthodox Hindus have been sprinkling themselves from little jars, but any resulting purification has been strictly spiritual. Gandhi in his autobiography mentions a Brahmin who 'would pour water over himself but never wash'. One has only to see and smell the Ganges to know that the mass immersions in that holy river at Benares have no connection with hygiene.

In Lapland the annual religious bath involved—and for all the research that we care to undertake may still involve—three duckings through a hole in the river ice. Newborn Russian babies were similarly treated in the frozen Neva, a practice that we may safely assert is now discouraged.

Water, preferably blessed beforehand, is the most common liquid for ablution, but such alternatives as cow's urine or blood have been used. The liquid must run off properly to carry away the evil. Demosthenes speaks of 'purifying the initiated and wiping them clean *with* (not *from*) mud and pitch'. Even holiness can be contagious, and render common things dangerously deo-active until ritually decontaminated. The Hebrew priest had to wash *after* handling the sacred book as well as before. In the strictest ecclesiastical sense ablution is the ritual washing of the priest's fingers *after* Holy Communion. Similar customs have been universal. A recent B.B.C. broadcast tells us that the fishermen of Milford Haven avoid washing when the catches are good, for fear of 'washing their luck away'.

Bathing has been found useful in sympathetic magic—to produce rain from heaven. It features also in the ritual of chivalry.

The story linking the Order of the Garter with a particular lady's garter (*hon y soit qui mal y pense*) has doubtful authority, but a 'Knight of the Bath' does indeed mean—as many a new K.B. has been dismayed to learn—a Knight of the Bathtub. The initiation formerly involved a ritual believed to have been instituted by Henry IV at his coronation in 1399. The process began by putting the candidate under the care of two 'esquires of honour grave and well seen in courtship and nurture and also in feats of chivalry'. Under their direction, a barber shaved him and cut his hair. He was then led to a bath 'hung within and without with linen and rich cloths'. They undressed him and put him in. Hot water is not mentioned. While he was in the bath, two 'ancient and grave knights' came to 'inform, instruct and counsel him touching the Order and feats of chivalry', after which they poured more water over him and retired. He was taken wet from the bath and put into a plain bed without hangings, where he stayed until dry, when the esquires dressed him in a white shirt and 'a robe of russet with long sleeves having a hood thereto like unto that of a hermit'. The ancient knights reappeared and led him to the chapel, the esquires going before them 'sporting and dancing, with the minstrels making melody'. After being served with wines and spices, they left him with the esquires, the priest, the chandler and the watch, to keep the vigil of arms until sunrise; the accolade came in the morning. The ceremony is simpler now.

Public baths, forgotten since the Roman occupation, came back to favour as the returning Crusaders taught the merits of the 'Turkish bath'. The Bathmen's Company used the sign of the Turk's Head, symbolising this importation from the East; there was a bath establishment in St. James's Street in the eighteenth century still known as The Turk's Head. This 'Turkish' bath was essentially a communal affair; Islam never accepted the tub bath in which, it was objected, a man merely soaked in his own dirty water; it was apparently quite acceptable to soak in other people's. Under Richard II the bath houses or 'stews' of London were owned by William Walworth, the Mayor, and there were no less than 18 in Southwark. Bankside, the Southwark waterfront, came to be known as Stewsbank, and even the lane leading from Upper Thames Street to the landing place opposite Southwark became Stew Lane. The 'stovers', who were members of the Barbers' Company, did such further offices as hair-cutting, shaving, cupping, blood-letting and minor operations. Boys ran through the streets announcing when the water was hot. The abundant steam generated in bread-making was sometimes used for 'stewing', much as the spare hot water from Battersea power station today serves the flats on the opposite side of the river. All that was needed was a pipe from the oven and room for the company. The Bathmen's Guild was in angry conflict with the Bakers' Guild over this infringement of trade.

58

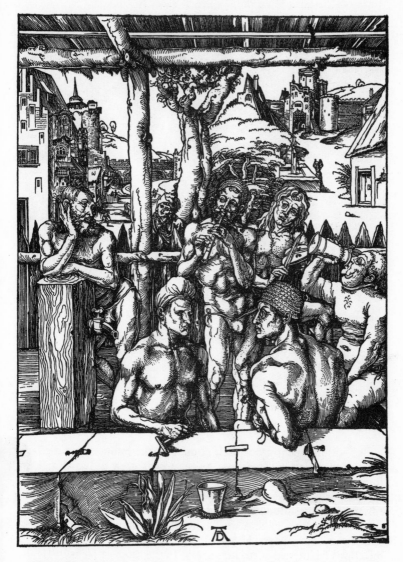

Dürer illustrates several scenes in the stews; in his native Nurem-berg the *Meistersingers* held singing practices there, and even had special bathing songs:

> Wine inside and water out,
> And merry we will be.

The Women's Bath: Dürer, 1496

But within a century of their coming in the public stews died out
again, for three main reasons. As the towns grew outward and the
forests dwindled, the source of wood for heating water was pushed
further away, and 'sea cole' was yet to come cheaply from Newcastle.
Immoral business done on the side led to objections from the church:
the words 'stews', 'bagnio' and 'bordello' took on another meaning,
which they have kept. Above all, the spread of plague and other
infections frightened customers away. In the reign of Henry VIII the
stews were closed by ordinance, and stayed closed or very strictly
regulated for nearly 150 years. Stow quotes one of these regulations:

Also, if there be any House wherein is kept and holden any Hot-house
or Sweating-house, for Ease and Health of Men, to which be resorting or
conversant any Strumpets, or Women of Evil Name or Fame. Or if there

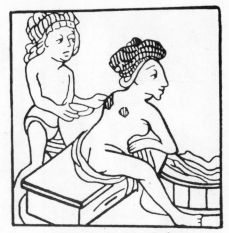

Bathman Cupping a Lady

be any Hot-house or Sweating ordained for women, to the which is any common recourse of young Men, or other Persons of evil Fame and Suspect Conditions. Also, if there be any such Persons that keep or hold any such Hot-houses, either for Men or Women, and have found no surety to the Chamberlain for their good and honest Behaviour, according to the laws of the City, and lodge any manner of person by Night, contrary to the Ordinance thereof made, by the which he or they shall forfeit 20£ to the Chamber, if they do the contrary.

They were having the same trouble in France, where *Rue des Étuves* is still a common street-name; the neighbours of the stews in Lyons complained of '*grands esclandres, tumultes, batteries, multres (meurtres), maulvais exemples et plusieurs autres dangiers et inconveniens*' and the stews were demolished by Francis I in 1538. The decay of the communal bath was so complete that we shall find it reintroduced into England in the seventeenth century as a new foreign luxury, and not fully revived until the late eighteenth century, by which time few cared to be clean, and bathing was a curative rather than a cleansing process.

The Stews: Cupping in Progress

Mediaeval London had many sources of water. A pious if simple friar once bade his hearers observe how Providence in its goodness almost always caused a river to run through any large town, to the great benefit of the inhabitants. London was triply blessed by this wise provision: Stow records that for 200 years after the Norman conquest water was drawn from three rivers, the Thames, the Fleet and the Walbrook; thereafter it became undrinkable. William FitzStephen in the twelfth century speaks of the then suburbs as having 'most excellent wells, whose waters are sweet, wholesome and clear'. These 'wells' were natural springs, some of them holy, such as Holywell and Clerkenwell. From early times there were a surprising number of artificial wells: when Watling Street was being cleared after the Great Fire, nine were found in a row serving nine adjoining houses. The first piped supply came in 1237 when Gilbert de Sandford, at the request of the King, granted to the city all the springs and waters in his fief of Tyburn at Mary le Bourne (Marylebone), with leave to bring them through his land by lead pipes to the city, and the right to repair the pipes and the tower where the waters were collected. Some water mains were, like those of the Romans, made of hollowed elm trunks, with the bark left on, fitted end-to-end with taper-and-socket joints bound with iron. The length and straight growth of elm, and its special resistance to water, made it most suitable. Wooden pipes are recorded as from 2 to 10 in. bore and from 10 to 22 ft. long. Machines driven by water-power were devised for boring them. Y's and T's were sometimes made from natural forks. Lead was often used, but earthenware pipes were less common.

There were many 'Cocks' or 'Conduits' offering running water. ('Conduit' could mean the main pipe or channel, or the conduit-head where the public drew water, or the building housing the conduit-head.) The Great Conduit of Chepe (Cheapside) was a stone building housing a stone basin and later a lead cistern, fed by the pipe from Tyburn, a distance of $3\frac{1}{2}$ miles. Yearly water-rates of from 5s. to 6s. 8d. were levied on the houses served. In the fourteenth century the houses in Fleet Street were often flooded by bursts in the main, which was by then over a hundred years old. In 1378 this pipe was extended to Cornhill, where another conduit-house was made economically by adapting a little old prison called The Tun.

The Palace of Westminster had water piped from the West Bourne at Paddington, probably since 1233 when Master William the *conductarius* or conduit-maker was sent 'to bring water into our court of Westminster in accordance with what we have told him'. The pipe was 'not thicker than a quill'. There was plumbing trouble there in 1373, when we find five workmen examining the pipes of the conduit 'which was broken and it was not known where, and water did not come to Westminster', and next week they are back again 'because the water was lacking in the hall and the kitchens and did not come through'. All too clearly we hear the arguments and excuses, and the criticisms of the British workman, across the centuries.

The citizens of Westminster were granted the use of the overflow from the Palace system, just as the monks had shared their surplus, but in Dublin this arrangement was reversed: in the fifteenth century a water-supply was brought in 'at the cost of the citizens', but the town granted the use of water to the monastery. The main pipe in Dublin was specified to be 5 in. diameter, but inside the houses the pipes were to be 'not larger than a man's little finger', and later 'of the size of a goose quill'. Subsidiary pipes became known as quills.

In Hull, the thirteenth-century supply proving inadequate, and having even to be supplemented by bringing water in casks by boat across the Humber, water was drawn from outlying springs. The villagers whose supplies were thus reduced retaliated by putting dead animals into the channels. A scheme to use the Anlaby Springs was carried through only after some of its more voluble objectors had been hanged at York.

In November 1478, when naughty William Campion unlawfully tapped the conduit to serve his house in Fleet Street, the punishment fitted the crime: he was led through the wintry streets on horseback with a vessell like unto a conduyt full of water uppon his hede, the same water rennyng by smale pipes oute of the same vessell, and that when the water is wasted newe water to be put in the saide vessell again.

In Elizabethan times it was forbidden to use the conduits during church service, and if a tankard or pail were left there at that time the churchwardens might impound it until a fine of 4d. was paid into the poor-box. To celebrate occasions such as victories or coronations, the conduits sometimes ran with wine. This did not demand quite the liberal flow that one would like to imagine: the water was turned off and a thin subsidiary pipe supplied from a wine-barrel was inserted in the outlet. Most of the city conduits were destroyed in the Great Fire, but several were rebuilt, and some survived in use into the nineteenth century.

Mediaeval Water Carrier

A water wheel was built by Peter Morice, a Dutchman, in 1582 in one of the arches of London Bridge, working a pump to convey Thames water to the city. This was sufficiently powerful to throw a jet over the steeple of St. Magnus' church. We are told elsewhere that lead pipes were carried over the steeple, but unless there was a storage tank in the steeple (which was not Wren's later one, still standing, in which there would have been more room for it), there seems no point in this. No tank in the steeple is described, but there was later a tank on a specially-built tower, for which the jet over the steeple may have been a test. The antiquary Abraham Fleming notes that

the waste of the first main pipe ran first this year, 1582, on Christmas Eve. This main being brought up at the Citie 'change into a standard, and divided into several spouts, ran four waies, plentifullie serving the use of the inhabitants that will fetch the same into their houses. A great commoditie to the Citie.

Eventually five arches of the bridge were used for water-wheels; boatmen complained of the obstruction, and water-carriers of the competition. The water works stood until 1822.

At Broken Wharf in 1594 one Bevis Bulmer built an 'engine' worked by horses to supply Cheapside and Fleet Street, where the old conduit must have become inadequate, with unpurified Thames water. But this had long been undrinkable, and London like ancient Rome had to seek water from a distant source high enough to supply it by gravity. In 1613 Sir Hugh Myddleton brought water to London by the New River, an artificial channel 38 miles long from Chadwell in Hertfordshire, to the New River Head at Sadler's Wells. (The address of the Metropolitan Water Board's head office and laboratories is still 'New River Head'.) From Sadler's Wells ran wooden mains 'serving the highest parts of London in their lower rooms, and the lower parts of London in their higher rooms'. Six hundred men had worked for 5 years. No doubt the open channel was useful for bathing, clothes-washing and rubbish-disposal, but even so the New River was cleaner than the Thames. James I took a half-share in the scheme, and won the distinction of being the first person to fall in to the New River. A typical lease granted by what might be called the firm of James and Myddleton was for

a pipe or quill of half an inch bore, for the service of their yard and kitchine, by means of tooe of the smallest swan-necked cockes, in consideration of the yearly sum of 26s. 8d.

Until the seventeenth century, such private water services were few. Lord Cobham asked for a 'quill' of water from Ludgate Conduit in 1592, but this was refused. The pipe to Essex House was cut off by order of the Lord Mayor in 1608, as was Lord Burghley's also, because of wasteful habits. From the seventeenth to the early nineteenth century, householders were normally allowed to fill their storage tanks only at set hours.

The Worshipful Company of Plumbers, 'of a large and very memorable antiquity, remaining a Fellowship or Brotherhood by the name of Plumbers', was incorporated by Letters Patent in the ninth year of King James I. The Charter, making it unlawful for others to 'use the art and mystery', is still to be seen at the Guildhall Library, likewise the ancient Book of Ordinances, 38 Edward III, 1365, and 11 Henry VIII, 1520, testifying to the antiquity of the craft. Plumbers' Hall was in Bush Lane, on the site of Cannon Street station. The arms of the Company are blazoned

Or, on a chevron sable between a cross staff fesseways of the last, enclosed by two plummets (plumb lines) azure, all in chief, and a level reversed in base of the second, two soldering irons in saltire, between, a cutting knife on the dexter, and a shave hook on the sinister, argent.

VI

Dirty Days

CLEMENT VII, who was Pope from 1523 to 1534, was a Medici and a man of taste. His *stufetta* (or little stew, or bathroom) in the Castle of St. Angelo in Rome is still to be seen, with its marble bath, hot and cold water taps, hot air circulated in the old Roman manner behind the walls, and frescoes by Girolamo Romanino, a pupil of Giorgione. The Pope still shepherded pre-Reformation England into paths of godliness, but not, alas, into those of a like cleanliness. With the sixteenth century England went into a dirty decline. Erasmus, in a letter to Cardinal Wolsey's physician written about 1530, describes the English house with its rush-strewn clay floors deep in refuse, in language which would not lose its flavour if translated:

Tum sola fere sunt argilla, tum scirpis palustribus, qui subinde sic renovantur, ut fundamentum maneat aliquoties, annos viginti, sub se fovens sputa, vomitus, mictum canum et hominum, projectam cervisiam, et piscium reliquias, aliasque sordes non nominandas. Hinc mutato coelo vapor quidam exhalatur, mea sententia minime salubris humano corpori.

In short, the floor needed sweeping. Meals however still began with the ceremony of hand-washing. Not until the seventeenth century was the general use of forks to make this troublesome business unnecessary.

The mediaeval type of garderobe was no longer built, but not because any better method had come in. The 'close stool' was the chief substitute for the garderobe; cosier for the user but hard on the servants. Some of the royal seats were fit for kings—at least to the eye. Louis XI of France had a *retraict* in his room, an iron framework covered with curtains to screen the stool, and wisely allowed regular purchases of *coq mente* (tansy) and other herbs to sweeten its air. Some $15\frac{1}{2}$ ells of green damask, at a cost of £52 2s. 6d., were used in making another such pavilion for King James V of

Pope Clement VII's Bathroom in the Castle of St. Angelo, Rome, c. 1530

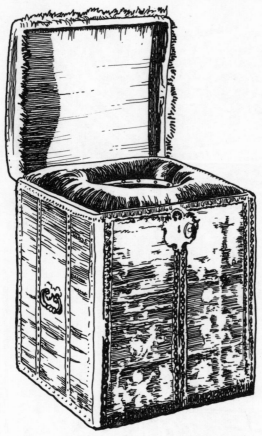

Royal Close Stool at Hampton Court,
c. 1600

Scotland's 'Stool of Ease'. Many close stools were made by the
Royal Coffer Makers in London, an appointment long held by the
family of Green or Greene, prominent members of the Leather-
sellers' Company, whose names appear often in the Royal Household
Accounts. A stool made in 1547 'for the use of the kynges mageste'
was covered with black velvet and garnished with ribbons, fringes
and 2,000 gilt nails. The seat and arms were covered with white
'fuschan' filled with down. Made like a box with a lid, this stool was
supplied with two leather cases lined with black cotton and fitted
with straps, one case for the stool and one for the bowl and 'sesstornes' or cisterns—for this seems to have been a portable *water*

closet, and may well have accompanied Henry VIII on his travels. A stool on which Elizabeth I or James I, or both in succession, may have sat, is still at Hampton Court. It is covered in crimson velvet, likewise bound with lace secured by gilt nails, and has carrying handles. The seat is padded with velvet. The lid may be locked against illicit use. There are no 'sesstornes', but a pot.

An end of such unsavoury devices might well have come as early as 1596, when Sir John Harington, a godson of Queen Elizabeth, wrote his *Metamorphosis of Ajax: A Cloacinean Satire* (the pun will escape those who do not know what 'a jakes' was), describing a valve water closet of his invention, erected at Kelston near Bath. It has a seat with a pan, a cistern above (in which are shown fish swimming, but only to indicate the water), an overflow pipe, a flushing pipe, a valve or 'stopple' and a waste with a water-seal. He illustrates it with 'Plan Plots of a Privy in Perfection' and fully details its construction, cost and maintenance.

To M.E.S., ESQUIRE.

SIR,

My master having expressly commanded me to finish a strange discourse that he had written to you, called the Metamorpho-sis of Ajax, by setting certain pictures thereto . . . Wherefore now, seriously and in good sadness, to instruct you and all gentlemen of worship, how to reform all unsavoury places, whether they be caused by privies or sinks, or such like (for the annoyance coming all of like causes, the remedies need not be much unlike) this shall you do.

If that which follows offend the reader, he may turn over a leaf or two, or but smell to his sweet gloves, and then the savour will never offend him.

AN ANATOMY.

In the privy that annoys you, first cause a cistern, containing a barrel, or upward, to be placed either in the room or above it, from whence the water may, by a small pipe of lead of an inch be conveyed under the seat in the hinder part thereof (but quite out of sight); to which pipe you must have a cock or a washer, to yield water with some pretty strength when you would let it in.

The cistern in the first plot is figured at the letter A; and so likewise in the second plot. The small pipe in the first plot at D, in the second E; but it ought to lie out of sight.

Next make a vessel of an oval form, as broad at the bottom as at the top; two feet deep, one foot broad, sixteen inches long; place this very close to your seat, like the pot of a close-stool; let the oval incline to the right hand.

The vessel is expressed in the first plot H.M.N., in the second H.K.

71

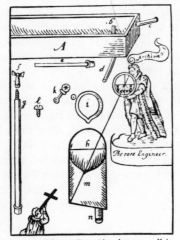 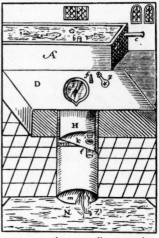

This is Don Ajax house...all in sunder; that a workman may see what he hath to do

...the same, all put together, that the workman may see if it be well

Harington's Watercloset, 1596. From his *Metamorphosis of Ajax*

This vessel may be brick, stone or lead; but whatsoever it is, it should have a current of three inches to the back part of it (where a sluice of brass must stand); the bottom and sides all smooth, and dressed with pitch, rosin and wax; which will keep it from tainting with the urine.

The current is expressed in the second plot K.

In the lowest part of the vessel which will be on the right hand, you must fasten the sluice or washer of brass, with solder or cement; the concavity, or hollow thereof, must be two inches and a half.

A special note.

In the second plot, I.L.

To the washers stopple must be a stem of iron as big as a curtain rod; strong and even, and perpendicular, with a strong screw at the top of it; to which you must have a hollow key with a worm to fit that screw.

In the first plot G.F., in the second F. and I.

This screw must, when the sluice is down, appear through the plank not above a straw's breadth on the right hand; and being duly placed, it will stand about three or four inches wide of the midst of the back of your seat.

In the first plot between G.I.

Item, That children and busy folk disorder it not, or open the sluice with putting in their hands without a key, you should have a little button or scallop shell, to bind it down with a vice pin, so as without the key it will not be opened.

This shows in the first plot K.L., in the second G.; such are in the backside of watches.

WELL DONE AND ORDERLY KEPT

These things thus placed, all about your vessel and elsewhere, must be passing close plastered with good lime and hair, that no air come up from the vault, but only at your sluice, which stands closed stopped; and it must be left, after it is voided, half a foot deep in clean water.

Else all is in vain.

If water be plenty, the oftener it is used, and opened, the sweeter; but if it be scant, once a day is enough, for a need, though twenty persons should use it.

If the water will not run to your cistern, you may with a force of twenty shillings, and a pipe of eighteen pence the yard, force it from the lowest part of your house to the highest.

But now on the other side behold the Anatomy.

Here are the parts set down, with a rate of the prices; that a builder may guess what he hath to pay.

		s.	d.
A,	The Cistern: stone or brick. Price	6	8
b, d, e	the pipe that comes from the cistern with a stopple to the washer	3	6
c,	a waste pipe	1	0
f, g	the item of the great stopple with a key to it	1	6
h,	the form of the upper brim of the vessel or stool pot		
m,	the stool pot of stone	8	0
n,	the great brass sluice, to which if three inches current to send it down a gallop into the JAX	10	0
i,	the seat, with a peak devant for elbow room		

The whole charge thirty shillings and eight pence: yet a mason of my masters was offered thirty pounds for the like.

. . . And this being well done, and orderly kept, your worst privy may be as sweet as your best chamber.

*   *   *   *

Little do parents and nurses know what problems can torment the infant mind, over some baldly-stated but unexplained truth. God, we were told in the nursery, is everywhere. Was He, we wondered, in *every* room in the house? We might even ask this—not specifying the ungodly room we had in mind—and be assured by an inattentive grown-up that it was so. We worried over this. To press the innocent enquiry even with a seven-year-old sister was to be shunned as a blasphemer. Had Harington's book been to hand (an unlikely thing) this little problem of nursery theology would have been properly cleared up:

73

Sprinto non spinto. More feard than hurt.

A godly father, sitting on a draught
To do as need and nature hath us taught,
Mumbled (as was his manner) certain prayers,
    And unto him the devil straight repairs,
And boldly to revile him he begins,
    Alleging that such prayers were deadly sins
And that he shewed he was devoid of grace
    To speak to God from so unmeet a place.

The reverent man, though at the first dismayed,
    Yet strong in faith, to Satan thus he said:
Thou damned spirit, wicked, false and lying,
    Despairing thine own good, and ours envying,
Each take his due, and me thou canst not hurt,
    To God my prayer I meant, to thee the dirt.
Pure prayer ascends to Him that high doth sit,
    Down falls the filth, for fiends of hell more fit.

74

Sir John's invention embodies all the features of the valve closet, but although it was copied at the Queen's Palace at Richmond, its coming into general use, re-invented, was delayed for nearly 200 years.

He urges a daily cleansing of the whole body—'love you to be cleane and well apparelled, for from our cradles let us abhor uncleanness, which neither nature or reason can endure'. In 1598 there were 'bathing-rooms' attached to his godmother's Royal Apartments at Windsor Castle, one of them 'wainscotted with looking-glass', perhaps the room where Queen Elizabeth took a bath once a month 'whether she need it or no'. But with the dissolution of the monasteries their good example of a cleanly regime was lost. We may boast in many ways of the Elizabethans, but we find few references to bathing or washing in Shakespeare. We may boast also of the Age of Elegance, but not of its cleanliness; these two periods mark the beginning and end of two rather insanitary centuries.

Lord Bacon, writing in 1638, does not include soap among his many ingredients for a bath:

First, before bathing, rub and anoint the Body with Oyle, and Salves, that the Bath's moistening heate and virtue may penetrate into the Body, and not the liquor's watery part: then sit 2 houres in the Bath; after Bathing wrap the Body in a seare-cloth made of Masticke, Myrrh, Pomander and Saffron, for staying the perspiration or breathing of the pores, until the softening of the Body, having layne thus in seare-cloth 24 houres, bee growne solid and hard. Lastly, with an oyntment of Oyle, Salt and Saffron, the seare-cloth being taken off, anoint the Body.

Sir John Harington (1561–1612)

75

Allowing half an hour for undressing and dressing, and another half an hour for the rubbings and anointings, Bacon's bath could be completed in 27 hours flat. Such a drill would hardly have appealed to Charles II and his court, who spent the summer of 1665 in Oxford, and were berated in the diary of the Oxford antiquary Anthony à Wood:

Though they were neat and gay in their apparell, yet they were very nasty and beastly, leaving at their departure their excrements in every corner, in chimneys, studies, colehouses, cellers. Rude, rough, whoremongers; vaine, empty, careless.

The purpose of this visit was to avoid the Plague; not, as might be thought, to spread it. It must have been other delights than bathing that lured the Merry Monarch to Bath, Epsom and Tunbridge Wells.

Pepys was a little more particular: he had 'a very fine close stool' —in his drawing-room. He also had a cesspit in the cellar, and describes how its contents had to be brought up through the house. He kept his diary for nine years, but only once mentions his wife having a bath:

My wife busy in going with her woman to the hot house to bathe herself, after her long being within doors in the dirt, so that she now pretends to a resolution of being hereafter very clean. How long it will hold I can guess.

In *English Social History* G. M. Trevelyan paints a vivid picture of early morning in Edinburgh in Queen Anne's time:

Far overhead the windows opened, five, six or ten storeys in the air, and the close stools of Edinburgh discharged the collected filth of the last twenty-four hours into the street. It was good manners for those above to cry 'Gardy-loo!' (*Gardez l'eau*) before throwing. The returning roysterer cried back 'Haud yer han", and ran with humped shoulders, lucky if his vast and expensive full-bottomed wig was not put out of action by a cataract of filth. The ordure thus sent down lay in the broad High Street and in the deep, well-like closes and wynds around it making the night air horrible, until early in the morning it was perfunctorily cleared away by the City Guard. Only on a Sabbath morn it might not be touched, but lay there all day long, filling Scotland's capital with the savour of a mistaken piety.

Wherefore Defoe described the Scots as 'unwilling to live sweet and clean', though it is only fair to add that some thirty years later Hogarth was illustrating a similar scene in London. The streets of Edinburgh were perambulated by a man carrying a bucket and a great

NIGHT, by Hogarth, from Four Times of the Day, 1738

cloak, with the cry 'Wha wants me for a bawbee?' meaning 'Who wishes to hire the services of my bucket and shielding cloak for a halfpenny?' The Scots, growled Dr. Johnson, took good care of one end of a man, but not of the other. But Shanks of Barrhead was one day to retrieve their reputation handsomely throughout the world.

VII
Fecundating Springs

*The Hot Springs of Bath — Bladud's Leprosy Cured — Aquae
Sulis — Holinshed on Bath — Cross Bath for Leaprie — King's
Bath for Gentre — Water Reeketh Much — Queen Mary Con-
ceives — Dr. Pierce's Ailments — Pepys Remarks of Bath —
Celia Fiennes — Decorum and Indecorum — Tempting Amorous
Postures — Dr. Lucas Finds Bath Inelegant — Fanny Burney
Shocked — Buxton — Miraculous Waters — Harrogate — A
Tactless Analyst at Malvern — Tunbridge Waters Saponary and
Detersive — Sixty English Spas — 20 Pints at a Sitting —
London's Old Baths — Naked Butler in Garden — The Turkish
Hammams — Miss Pardoe's Views — Lady Mary Wortley
Montague's Feet — The London Hummums — Turkish and
Roman Baths Compared — Covent Garden Hummums — The
Duke's Bath — Casanova's Magnificent Debauch — Three
Shaves a Day — Earth Bathing — Bath Quacks — Dr. Johnson
Says No, Sir — Was England Merrie? — Improvements in the
Eighteenth Century — Dr. Cheyne's Advice — No Water Drunk
in London — Dr. Lucas' Essay on Waters — A Machine Wrought
by a Horse — London Waters Tested — Cats and Dogs in the
Reservoir — Blenheim and Chatsworth — The Alderman at
the Pump — Polynesian Savages — Dr. Domenicetti Junior —
Brighthelmstone*

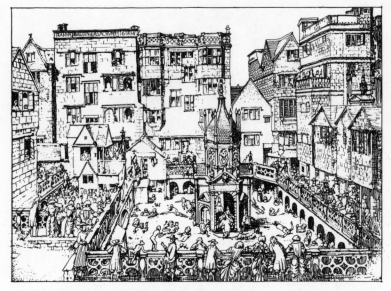

The King's Bath in 1675

LTHOUGH a bagnio scene with food, gossip and music (ladies only) is pictured in a pamphlet of James I, the communal bath did not much revive until towards the end of the century. The hot springs of Bath were an exception. These springs, lost since Roman times, had been rediscovered in the twelfth century. The archaeologists of that time explained that the springs had been found first in 863 B.C. by the British Prince Bladud, the father of King Lear. Bladud was that rather rare combination, a leper, a swineherd and a glider pilot. He had been banished as a leper, and so became a swineherd; his gliding is another story. His pigs were so improved in health by a chance immersion in the waters, that Bladud followed their lead, and was cured of his leprosy. He founded the city of Bath and dedicated the springs to Minerva. As evidence of the truth of this legend, we have the inscription on the statue of Bladud set up in 1699. Some may prefer the simpler legend, favoured by Geoffrey of Monmouth, that Bladud gave the Bath waters their power by the mere exercise of magic. The Roman name for Bath was *Aquae Sulis*, the Waters of Minerva. In Saxon times the place had the pleasing name of Hat Bathun.

80

Holinshed in the sixteenth century describes the Cross Bath as

much frequented by such as are diseased with leaprie, pockes, scabs and great aches . . . it is worthilie called the hot bath, for at the first coming, men thinke that it would scale their flesh, and lose it from the bone, but after a season . . . it is more tolerable and easie to be borne.

The King's Bath is 'verie faire and large' and has 32 arches for men and women to stand in separately,

who being of the gentre for the most part, doo resort thither indifferentlie, but not in such lascivious sort as unto older baths and hot houses of the maine whereof some write more a great deale than modestie should reveale and honestie performe.

The water, he says, is

most like to a deep blewe, and reeketh much after the manner of a seething pot, commonlie yielding somewhat a sulphurous taste, and very unpleasant savour.

The baths were closed around noon and midnight for the very necessary business of changing the water. They had lately been

very much repared and garnished with sundrie curious pieces of workmanship.

The waters of Bath won some publicity when Mary the Queen of James II, having heard of their wonder-working powers in cases of barrenness, spent some time in the Cross Bath, and had the satisfaction to find that 'fame had not exaggerated in her praises of these fecundating springs'. The Queen conceived, and as a memorial to the happy event an appropriate pillar was erected in the centre of the bath.

Dr. Pierce, who practised at Bath from 1672 to 1697, seems, as they say, to have 'enjoyed very poor health' until he took the waters. At the age of ten he had a dropsie, an ascites and an anasarcha together; at twelve the smallpox; at fourteen a severe tertian ague for six or seven weeks; then headaches, defluctions of rheumes to teeth, jaws and palate as well as to the glandules of the throat; at twenty-one, 'breeding the measles', he bled at the nose for two days and they despaired of his life; at thirty, living 'near the moors and mashy country' he took an epidemick fever which determined in a quartan ague; a swelling of hemeroid veins and scorbutical symptomes; in which state he moved to Bath, took the waters, recovered promptly and lived healthily to the age of seventy-five.

Pepys remarks of Bath 'methinks it cannot be clean to go so many bodies together into the same water'. Celia Fiennes was there about 1687 and describes how the ladies, wearing special gowns of stiff yellow canvas, entered and left the water through a wooden door so that they were not seen until submerged to their necks. At a later date, conduct seems to have been less decorous:

Here is perform'd all the Wanton Dalliances imaginable; celebrated Beauties, Panting Breasts, and Curious Shapes, almost Expos'd to Publick View: Languishing eyes, Darting Killing Glances, Tempting Amorous Postures, attended by soft Musick, enough to provoke a *Vestal* to forbidden Pleasure, captivate a Saint, and charm a *Jove*: Here was also different Sexes, from *Quality* to the Honourable *Knights*, Country *Put* and City *Madams* . . . the ladies with their floating *Jappan* Bowles, freighted with Confectionary, Knick-knacks, Essences and Perfumes, Wade about like Neptun's Courtiers, supplying (suppling?) their Industrious Joynts. The Vigorous Sparks, presenting them with several Antick Postures, as Sailing on their Backs, then Embracing the Element, sink in Rapture.

But one lady visitor whose words ring more truly was unable to agree that one's bedraggled appearance in the bath went well with Tempting Amorous Postures, and a Mr. Hammond in his guide book said that the spectacle of the mixed half-naked crowd put him in mind of the Resurrection. Dr. Lucas in 1756 complains that there are now no places for undressing; the bather must be carried undressed from his lodgings; must take his chance for the temperature of the air as well as that of the bath; that he must bathe in sight of spectators who 'divert themselves as at a bull- or bear-baiting'. In the bath he wears a tight canvas jacket and drawers, and a cap on his head, which attire 'must frustrate the intention and the end of bathing'. He is sent home

Comforts of Bath: Rowlandson, 1798

wrapped in blankets. He finds it all 'extremely inelegant'. But with him at the 'shamefully rude and barbarous baths' was Mr. John Wood the architect, who with Nash and Allen was to transform Bath into a city of Palladian magnificence. Fanny Burney in 1780 was shocked at the public exhibition of ladies in the bath; 'it is true, their heads are covered with bonnets; but the very idea of being seen in such a situation by whoever pleases to look, is indelicate'. Another complaint was of the rubbish dropped in the baths, especially the nut-shells, cherry-stones and plum-stones, until daily cleaning was adopted.

Buxton, like Bath, goes back to Roman times, though no remains survive. The history of the town as a spa dates from about 1570, when in *The Benefit of the Auncient Bathes of Buckstones* Dr. Jones claimed that although the water of Bath was hotter,

by reason whereof, it attracteth and dissolveth more speedily; but Buckstone's more sweetly, more delicately, more finely, more daintily, and more temperately.

The scale of charges at Buxton was steeply graded by a means test, so that a yeoman paid 1s., a duke £3 10s. and an archbishop £5. The power of spa and well waters was regarded as miraculous rather than medicinal. Dr. Jones expected the bather to pray, either at home or at the edge of the bath, before entering the water. The Bishop of Bath and Wells wrote special prayers for bathers, and warned them that 'God must first *heal the Waters*, before they can have any virtue to heal you'. Although at the Reformation Thomas Cromwell's agent locked up and sealed the sources, belief in their holy powers was not destroyed, and the holy wells are still venerated by the Derbyshire dalesmen at their annual 'well-dressing' ceremonies.

At Harrogate, where the flow was less abundant, the visitor was treated in a hot solo tub; a versifier describes it:

> Astonished I saw when I came to my doffing
> A tub of hot water made just like a coffin,
> In which the good woman who attended the bath
> Declar'd I must lie down as straight as a lath,
> Just keeping my face above water, that so
> I might better inhale the fine fumes from below.

Another early spa was Malvern, but a tactless analyst who examined the waters declared them quite pure—not quite what is required of medicinal waters:

> The Malvern water, says Doctor Wall,
> Is famed for containing just nothing at all.

THE GROWTH OF A SPA

Plombière-les-Bains is one of several 'health springs' in the Vosges (it is near to the pleasingly-named spa Bains-les-Bains). Its development is shown in these wood-cuts of different dates. Note the large number of 'pub signs' in the second picture, dated 1553. The bathwater seems a little more drinkable in the third picture, with its fountain, than in the second

The Tunbridge waters, said Dr. Madan in the seventeenth century,

with their saponary and detersive quality, clean the whole microcosm or body of man from all feculency and impurities. No remedy is more effectual in hypochondriacal and hysterick fits by suppressing the anathymiasis of ill vapours, and hindering damps to exhale to the head and heart.

Dr. Falconer of Bath, writing in the eighteenth century, said that there were already more than a thousand treatises on mineral waters. At least sixty English spas eventually flourished, but at most of them the flow was limited and the cure confined to drinking the waters, some of which

nothing but the most extraordinary fear of death, or the most singular insensibility to foulness in taste and smell, could ever have reconciled any human being to touching after the first drop,

and sometimes drinking them in prodigious quantities, up to 20 pints at a sitting.

The argument behind the doctors' belief in water cures was, that since there was no complaint for which Nature had not provided a suitable healing spring, analysis of the waters would show which diseases they were intended to cure. They found, for example, that the Scarborough waters cured 'Preternatural Thirst, All Sorts of Worms, and Disorders of the Stomach from Intemperance'; the Tunbridge Wells waters 'strengthened the Brain and Origin of the Nerves, and were good for Head-ach and Vertigo'; the Bath waters cured everything from 'Cold Humours and Hypochondriacal Flatulence' to 'the Longing of Maids to eat Chalk, Coals and the Like'.

Some of London's old public baths have been very long in use. The St. Agnes-le-Clair Baths in Tabernacle Square, Finsbury, may have been Roman; Roman coins were found there. The spring is said to have produced 10,000 gallons a day. Queen Elizabeth's Bath at Charing Cross, a square pool with surrounding steps where the bathers sat, under a brick vault, stood until 1831. Queen Anne's Bath, so called after she had used it, but much older, was in Long Acre. This bath was fed by a spring reputed sovereign against the rheumatism. Clerkenwell Cold Bath, near the top of Mount Pleasant in the Cold Bath Road, also with a curative spring, had a chair hung from the roof so that a weak patient could be lowered into the water, for which an extra 6d. was charged above the normal fee of 2s. The Peerless Pool in Baldwin Street, City Road, was big enough for swimming, and was so used by the scholars of Christ's Hospital. The Royal Bagnio, opened in 1679, stood in Roman Bath Street until 1876. The Floating Baths on the Thames off Somerset House were

Queen Elizabeth's Bath

Queen Anne's Bath

patronised by the Queen of William III. The old Roman Spring Bath in Strand Lane, off Fleet Street, was still in use in Dickens' day: David Copperfield had 'many a cold plunge' in it.

In America, one of the earliest exponents of the merits of the cold plunge was the butler employed by William Penn (1644–1718) the founder of Pennsylvania. This unhappy man was not only deaf, but was 'long vexed with wandering pains and aguish accessions'. At the height of these afflictions he leaped from his bed on a cold night, threw off his night shirt, jumped into cold water, ran naked round the garden, into the water again, twice more round the garden; then, taking 'a good swigg o brandy', back to bed—and needless to say had recovered both health and hearing by the morning.

Late seventeenth-century travellers, like the Crusaders, brought back tales of the Turkish *Hammams*. A Miss Pardoe, who had spent two and a half hours in one at Constantinople, declared that she saw none of the 'unnecessary and wanton exposures' objected to by Lady Mary Wortley Montague. Lady Mary was in any case a tainted witness: her horrid habits were known, and are illustrated by a story of a French lady who, expressing surprise at the grubbiness of Lady Mary's hands, was told 'Ah, madame, if you were to see my feet!' As the 'Hummums' the Turkish baths were imitated in London. Ned Ward, in *The London Spy* of 1699, describes their stifling air, 'hot as a pastrycook's oven'. The Turkish bath was by no means the same as the Roman bath. Giedion explains the difference:

> The palestra and its gymnastic games, together with the swimming pool in the frigidarium, disappear . . . the high-windowed, light-flooded tepidarium gives way to cupolas sparsely pierced by the glow of coloured bullions, or to stalactite cupolas in the smaller rooms. Half light, quiescence, seclusion from the outside world are preferred . . . the active bather of the classical world yields to the passive repose of the oriental. A refined technique for loosening, cracking the joints, and a shampoo massage with special penetrative power supplant athletic sports.

Admission at one Hummum cost no less than 8*s*., apart from such extras as private rooms, wine, and sweet herbs for the bath-water. The hummums in Covent Garden opened in 1708 and flourished for over 150 years. At the Duke's Bath in Long Acre an oval bath hall was paved with marble, and four adjacent chambers had varying temperatures. The pleasures offered were thought cheap by Casanova, who was in London about 1765 and says

> I also visited the bagnios where a rich man can sup, bathe and sleep with a fashionable courtesan, of which species there are many in London. It makes a magnificent debauch and only costs six guineas.

88

Jost Amann's sixteenth-century woodcut purporting to show a Turkish lady's bathing costume is probably based less on authentic information than on erotic fancy

It was during this stay in London that Casanova heard his friend Lord Pembroke telling his valet to shave him:

'But,' said I, 'there's not a trace of beard on your face.'
'There never is,' said he; 'I get myself shaved three times a day.'
'Three times?'
'Yes, when I change my shirt I wash my hands; when I wash my hands I have to wash my face, and the proper way to wash a man's face is with a razor.'

This was the period of Dr. Graham's Temple of Hygieia and his often-described Celestial Bed (£100 per night, with the aphrodisiac effects of the magnets and other devices almost guaranteed). Dr. Graham preached the virtues of earth-bathing, publicly demonstrated when the Doctor and a young lady were buried up to their necks; though they were lightly clad, their coiffures were elaborate, and a spectator remarked that they looked very like cabbages. Similar semi-magical practitioners were the Hydropathists, or 'Bath Quacks' to the sceptics. The most famous was Dr. Domenicetti of Cheyne Walk in Chelsea, who achieved the feat of writing a whole book about his treatment and its benefits, without giving away a single fact about his methods. Dr. Johnson was among the sceptics:

'There is nothing in all this boasted system. No, Sir; medicated baths can be no better than warm water: their only effect can be that of tepid moisture.' One of the company took the other side; he turned to the gentleman: 'Well, Sir, go to Domenicetti, and get thyself fumigated; but be sure that the stream be directed to thy head, for *that* is the *peccant part*.'

Buer in his *Health, Wealth and Population in the Early Days of the Industrial Revolution* upsets some common notions of the sanitary history of England. He rejects the popular picture of a Merrie England gradually declining into dirty ways as the towns grow, and then plunged rapidly into squalor by the industrial revolution. In many ways the mediaeval towns were rural communities, with the backward sanitary habits of such. These habits might be fairly harmless in an isolated hamlet, but they could be fatal to health in a town. A town run on village lines he compares to an army, which soon suffers from dysentery and fevers if it settles down too long in one place. The primitive mediaeval sanitary regulations had no real popular approval behind them and were hardly enforced. Famine and pestilence were never far distant in that Golden Age to which some look back. There was no Mayhew to record the horrid details. Was the thirteenth-century peasant healthier and happier than the nineteenth-century factory hand? No one can answer.

The Pump in Bishopsgate Street in 1814

These habits survived into later centuries. Insanitary towns are not a modern invention. There was not a steady downward trend. The seventeenth century was a period of civil war and disorder; many severe epidemics are recorded, including visitations of the Plague, but there is no proof that conditions were worse than in mediaeval London. By the eighteenth century improvement had begun. In the first half of the century England's population increased by one-fifth (by one million) but the rise in the death-rate was due more to cheap gin than to bad sanitation. From 1750 to 1800 London's population increased by one half, but the death-rate fell. Fewer persons died in the large London of 1797 than in the small London of 1697. Though the towns were growing, their conditions were becoming more healthy. Streets were being widened and paved, drains covered in, houses rebuilt in brick and at much lower densities. Steam pumps (1743) and iron pipes (1746) came in, to improve the water supply. Cheap cotton cloth appeared; cotton can be boiled without spoiling it; boiling kills lice. Cheap crockery and iron utensils also encouraged cleanliness. London's privies in 1752 were, according to one account, 'well regulated'. If properly carried out the sewage disposal methods of the time were not necessarily insanitary. Writing of Paris in 1787 Arthur Young found it vastly inferior to London in cleanliness, though not in social and intellectual life. In other towns, bad as many were, there was some slow improvement, as the Bills of Mortality show.

Bathing might be rare in the eighteenth century, but hygiene does not live by baths alone. Public sewers and water-closets were being installed. Hospitals improved. Medicine ceased to be a blend of alchemy and magic, and became a science: when in 1720 a woman in Godalming declared that she was giving birth to rabbits, several doctors, including the King's anatomist, believed her. Twenty years later no doctor could have been thus deceived. The scientific age had begun.

Dr. George Cheyne recommends bathing and washing in 1724:

I cannot forbear recommending *Cold-bathing*; and I cannot sufficiently admire, how it should ever have come into such *Disuse*, especially among *Christians* . . . frequent *washing* the Body in *Water*, cleanses the Mouths of the *Perspiratory Ducts* from that *Glutinous* Foulness that is continually falling upon them, from their own condensed *dewy Atmosphere*, whereby the *Perspiration* would soon be *obstructed*, and the *Party* languish . . . I should advise therefore, every one who can afford it, as regularly to have a *Cold Bath* at their House to wash their Bodies in, as a *Bason* to wash their hands; and, constantly, *two* or *three* Times a Week, *Summer* and *Winter*, to go into it. And those that cannot afford such *Conveniency*, as often as they can, to go into a *River* or *Living Pond*, to wash their Bodies . . . I cannot approve the *precipitant* Way of *jumping* in, or throwing the Head foremost into a *Cold Bath*; it gives too violent a Shock to Nature, and risques too much the *Bursting* some of the smaller Vessels. The Natural Way is, holding by the *Rope*, to walk down the Steps as fast as one can and when got to the *Bottom*, bending their *Hams* (as Women do when they Curt'sy low) to *shorten* their Length, so as to bring their Heads a good Way under *Water*, and then *popping* up again to take Breath . . .

He recommends also, to the Tender, Studious and Sedentary, the shaving of Face and Head (a wig being worn all day), and Washing, Scraping and Paring the Feet and Toes. A few drops of Compound Spirit of Lavender or Hungary Water may be used with the washing water, or even soap, which will cleanse the Mouths of the Perspiratory Ducts from Morphew and Scurf, and by encouraging Perspiration will give a full and free Vent to the Fumes on the Head and Brain.

SIMPLE WATERS

In 1726 a Swiss visitor to London, M. de Saussure, is astonished
at the amount of water used. He says that absolutely none is drunk;
that the lower classes, even the paupers, do not know what it is to
quench their thirst with water. Gin was certainly cheap ('Drunk for
1d., dead drunk for 2d., straw free'), but hardly a thirst-quencher;
beer must have been the mainstay. By 1756 Dr. Lucas in *An Essay on
Waters* could boast that London was served with 'simple waters' in
greater variety and abundance than any city in Europe—the word
'variety' suggests some impurities. There is not, he says, a considerable street in London which is not furnished with such plenty of
water, that not only the ordinary offices on the ground floor or
under it, but even the upper storeys of most houses are, or may be,
supplied from the common 'aqueducts' in the streets. This supply, he
adds, is one of the causes why our capital is the most healthful city
in the world. He mentions four main sources: the Thames, from
which water is drawn by machines at London Bridge, Chelsea and
York Buildings; New River; Hampstead Ponds from which water is
piped to town; and a large spring at the end of Rathbone Place whose
waters are raised by 'a machine wrought by a horse'. He tests the
waters and boils them dry to weigh the residue. The results are surprising: a gallon of Thames water yields 16½ grains, New River water
14½ grains, Hampstead water 90 grains, Rathbone Place water
100 grains. Such residues can not be the only factor in assessing
potability.

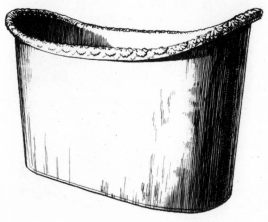

Tin Bath, eighteenth century

93

The water supply was intermittent: each house received water for two or three hours on about three days a week, when the tanks would be filled. Ball-valves came in about 1748, and saved the householder not only from having to remember to open his main tap, but from the sad consequences of forgetting to close it. With the supply so limited, waste was a serious matter; in 1770 the Bath authorities threatened to cut off Mr. Melmoth's water unless he disconnected his new water-closet. At York the quality of the water was such that most houses had large pots in which it was left to settle for a day or two before use. Manchester issued a proclamation in 1765 forbidding the drowning of cats and dogs, and the washing of dirty linen, in the Shute Hill reservoir.

The Pump in Cornhill

Country mansions sometimes had water systems of their own, the fountain court being for use as well as ornament. At Blenheim Palace the eastern archway contained a great cistern to receive water from the 'engine', from which it served the kitchen and offices by gravity.

As early as about 1700, Celia Fiennes saw at Chatsworth a 'bathing room' with blue and white marble walls, and a deep bath fitted with '2 locks to let in one hott, ye other cold water to attemper it as persons please'. This was a rare luxury. For those who were neither wealthy nor unhealthy, but could face a brisker wash than the Bason offered, there were countless pumps, as street- and place-names still attest. Alderman Boydell, the respected publisher, who became Lord Mayor of London, did not think it beneath his dignity to repair daily to one in Ironmonger Lane, on top of which he always placed his wig before dousing his head under the spout. But no readers of Captain Cook's account of the Tahitians, who invariably washed three times a day and kept their clothes extremely clean, seem to have felt any urge to introduce the habits of Polynesian savages to London.

A son of the famous Dr. Domenicetti, after seven years as an assistant at the Cheyne Walk establishment, transferred the business from the village of Chelsea into town, erecting his 'Apparatus of Arbitrarily heated and medicated WATER BATHS, partial PUMPS, vaporous and dry BATHS, internal and external moist and dry FUMIGATIONS, oleous, saponaceous, spiritous and dry FRICTIONS', in Panton Square, Haymarket, in 1779. He is as reticent as his father about the details of his Sweating Bed Chambers and of the 'many machines, pipes &c.' which 'compose and form a curious body of mechanism'. His machine for pumping on any part of the body (which would in later times have been termed a 'douche') is 'so conveniently constructed' that the flow may be varied from the smallest drops to 'a full quantity and strength of a fire engine'—fortunately for the patients, fire engines were less powerful in those days. He is critical of rival methods such as those used at Bath, where patients 'sent to be pumped on the head for the palsy' have, he says, lost their lives under the operation. As for those who prefer sea water to his special mixtures, he gives several proofs of the harmful nature of the coastal brine: it rusts iron, and so must be corrosive; water deprived of its salt by distillation will not putrefy, and so salt water must hasten putrefaction; sailors at sea have scurvy, and so salt must be the cause. Despite these frightful risks, sea bathing was coming steadily into fashion. A picture of Scarborough beach in 1735 shows gentlemen bathing, and at Margate in 1750 'Beale's Bathing Machines' dragged by horses took either sex into the water, which they could enter under the decent cover of a hood.

One of the earliest records of sea-bathing at Brighton (then Bright-helmstone) is in a letter written by the Rev. William Clarke, who stayed there in 1736. The practice flourished under Dr. Russell, who built himself a house there in 1754, just after publishing his *Dissertation Concerning the Use of Sea-water in Diseases of the Glands*. He advised cold weather for sea-bathing; Fanny Burney, Mrs. Thrale and the Misses Thrale bathed at Brighton before dawn, in November. Dr. Russell introduced a novelty: the drinking of sea-water. The Brighton brine was found so much more efficacious than that of other places that it was bottled and sent to London:

TO BE SOLD, at the Talbot Inn, Southwark, Sea-water from Bright-helmstone, in Sussex, took off the main Ocean by T. Swaine.

Dr. Awsiter advised his patients to mix their sea-water with milk, boiled with a little cream of tartar, then strained and cooled before downing the nauseous brew. There were fresh water springs too at Brighton.

The extraordinary fecundity of the sheep which drink this water gives the shepherds of this place an opportunity of extolling its prolific power.

Dr. Awsiter introduced another novelty: hot sea-water baths, which released Poisonous Humours through the pores. For ordinary sea bathing, a machine had to be hired, together with the services of a 'dipper', a muscular native who controlled the immersion. Dr. Johnson visited the Thrales at Brighton and bathed in the sea, perhaps hoping to cure the skin trouble for which Queen Anne had 'touched' him in vain in his boyhood. A not very tactful remark by his dipper is recorded: 'Why, Sir, you must have been a stout-hearted gentleman forty years ago!' Mixed bathing was not allowed, but with the aid of a spy-glass, as Rowlandson has shown, the well-wrapped beauties could be watched from afar:

> . . . on Brighton's sands the lovely Fair
> Smile at the rocks and hollow surges dare;
> When here so many Queens of Love we see
> Caress the waves and wanton in the sea . . .

But to the eighteenth-century eye only the kindest and prettiest of landscape or seascape was acceptable; high mountains or wide seas were awful and depressing, and at Brighton most of the new houses were built with their backs to the sea.

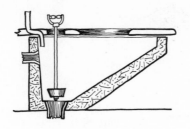

VIII

Lieux à l'Anglaise

Royal Palaces of France — Francis I — The Siamese Ambassador Impressed — Louis XIII — Submerged Cushions — Louis XIV's Pink Marble Octagon — River Bathing at Fontainebleau — Heating the Seine — 100 Bathrooms at Versailles — Du Barry — Marie Antoinette — Two Baths for One — Blondel — Chaises Percées — The Dauphin Commits a Nuisance — 264 Stools — The Pompadour — A Two Seater — Official Role of the Stool — 'Ladies' and 'Gentlemen' Invented — Voyage au Pays Bas — Mystères de Paris — The Good Name of England — Lieux à l'Anglaise — Queen Anne's Little Place — Carew's Pretty Machine — A Two Holer at Holkham — The London Art of Building — The Compendious View of All Trades — One Within at Woburn — A Complete Body of Architecture — Bog Houses — One W.C. at Kedleston — Embarrassing One at Osterley — Walpole Visits a Loose Family — The Eighteenth-Century W.C. — William Hawkins — The Pan Closet — Banner's Patent — The Plunger or Plug Closet — Cummings' Patent — Bramah's Patent — A Gulering Noise — Two Plumbers Stunned

I N the Royal palaces of France, magnificent bathrooms were built from time to time, but the delights of bathing seem to have palled more quickly than those of building, and they were constantly being dismantled and renewed. Francis I had hardly used his new *appartement de bains* at Fontainebleau when it was converted to other uses. A grander one of 1686 at the Louvre, with marble columns, a décor all of blue and gold, and a marble bath with running hot water, impressed the Siamese ambassador more than anything else in France, but this too had been demolished by 1722. Louis XIII was content for most of his lifetime with an ordinary wooden tub, though in his last days he agreed to a marble one. Grander than wood, marble offered a cold reception to the royal rump, and needed a proper *équipage de bain* consisting of submerged cushions and drapings trimmed with lace. Such outfits were popular as gifts in court circles.

For Louis XIV, as the incarnation of the Sun God, a really worthy receptacle was needed. In 1677 he had put in two new marble baths; in 1678 two more, with bronze ornaments; in 1679 he had brought the total up to six, but the suite was again enlarged, to take a monumental tub, a rose-pink marble octagon ten feet wide and three feet deep. The gilding of the ceiling alone cost 24,000 livres. When the Sun God deigned to bathe, in addition to the cushions and drapings and lace a great fabric pavilion was erected. Fortunately for the staff, he seldom did use it; he found the little bath in his bedroom cosier. Once again all the work was undone, and Louis XV gave the great bath to the Pompadour. Its removal was not easy. Even to move it on the level, on rollers, 22 men were needed. It was at first planned to demolish the massive vaulted floor, and get the bath away through the cellars, but it was realised in time that the window opening was just big enough, and with the help of ropes, capstans and a huge timber scaffold, up and out it went. Well may some seeds of the Revolution have been sown that day among the sweating workers. This bath, after a long spell of duty as a garden fountain, has lately been returned to Versailles and set up in the Orangery.

The court was cleaner than has generally been supposed; the contrast between the silk and lace finery and its malodorous wearers is easily and often drawn, but the slur is unjust. There were at least a hundred bathrooms at Versailles, where every important suite had one. It was in the later and lousier Age of Reason that they were dismantled. In one sense there was, *après moi*, no *déluge*.

River bathing had its devotees in the Court at Fontainebleau, where mixed parties went to the Seine, well wrapped in *grandes chemises de toile grise* down to their ankles. The Duchess of Burgundy, taking a dip at Marly, did things in style, with a leafy screen built along the bank and two pavilions with a magnificent banquet and the music of hautboys. If the river was too cold, boiling water was poured in to it. Let nobody say that this must have been ineffectual: poor Mme de Saint-Hérem allowed it to be overdone and was taken home to bed with sorry burns.

99

Du Barry's and Marie Antoinette's bathrooms are still on view. Marie Antoinette bathed daily, but was eccentric in that she used only one bath at a time. As Blondel, architect to Louis XV, explains in his *Maisons de Plaisance* of 1738, the bathroom of the time had two baths, one for washing and one for rinsing (on the principle of the modern double kitchen sink), thus answering for a while the oriental gibe at the occidental habit of soaking in one's own dirty water. This is why the French wrote always of the *salle de bains* in the plural. In mediaeval times we found one bath for two; now we find two baths for one.

Blondel shows a fine bathroom project, with two semicircular baths fitted into the corners. These have *impériales* (baldachinos or canopies) with silk curtains. Porcelain censers on sculptured pedestals flank the doorway. The taps and fittings are gilt, and all other plumbing is concealed. The walls are panelled in rare woods, but Blondel generally advises marble, both for *noblesse* and for *fraîcheur*.

The court was well equipped with *chaises percées*, which enjoyed the usual wealth of synonyms somehow inseparable from this sub-

The Fountain of Youth

ject, as *chaises d'affaires, chaires pertuisées, chayères de retrait, chaises nécessaires* or simply *selles*. The principle seems always to be, that once the term is generally understood, a new term must be invented to 'wrap it up'. Havard, in his *Dictionnaire de l'Ameublement*, avoids a choice between all these when he speaks simply of *ce meuble odorant*. Previous courts had not been well furnished in this respect, according to the behaviour of the Dauphin who on August 8, 1606, the very day on which an edict went forth forbidding any resident of the Palace of St. Germain to commit a nuisance within its confines, committed one against the wall of his own bedroom. An inventory of Louis XIV lists a total of 264 stools at Versailles. Of these, 208 were covered with *damas rouge, cramoisi ou bleu, de maroquin du Levant rouge, de moquette rouge, de velours rouge ou vert.* Sixty-six had concealing drawers or covers or were otherwise disguised. There was a round drum form known as a *Marseillaise*. Louis XV's smartest stool was in black lacquer with Japanese landscapes and birds in gold and coloured relief, with inlaid borders of mother-of-pearl, Chinese bronze fittings, red lacquer interior, and a padded seat in green velour. As Havard says, *on n'imagine rien de plus aimable*. The Pompadour was so delighted with the third and most splendid stool made for her by M. Migeon, that she got for him a pension worthy of a field-marshal. Another maker produced *une manière de chaise percée où l'on peut s'installer deux à la fois*.

The dignifying of the Royal Stool by these costly trappings had some point: it played an official role. Kings, princes and even generals treated it as a throne at which audiences could be granted. Lord Portland, when Ambassador to the Court of Louis XIV, was deemed highly honoured to be so received, and it was from this throne that Louis announced *ex cathedra* his coming marriage to Mme. de Maintenon.

Gradually the stool lost its place of honour, and was reduced to being hidden away inside another piece of furniture or disguised as something else. A model popular in France at the time of the Dutch wars was known as *Voyage au Pays Bas*, this being the title of the pile of large dummy volumes that hid the secret. The only variant title was *Mystères de Paris*: despite the scope for wit, it was important that the work should be recognisable for what it was. It would have been hard on the guest, ready for bed in his remote bedchamber, to find at the last moment that the books were real.

Perhaps the first mention in history of 'Ladies' and 'Gentlemen' in this connection is in the report of a great Ball in Paris in 1739, which tells, as of a remarkable innovation, that they had even taken the precaution of allotting *cabinets* with inscriptions over the doors, *Garderobes pour les femmes* and *Garderobes pour les hommes*, with chambermaids in the former and valets in the latter.

It is disconcerting to those who cherish the good name of England, to find it enshrined in the water-closets of France, where *anglaise* is defined as *surtout une sorte de garderobe*. Blondel in 1738 illustrates *cabinets d'aisance à l'anglaise* or *lieux à soupape* (valve closets). He says that these originated in England but that enquiries among his English friends reveal no knowledge of their existence in London. In 1750 the Paris dancer Mlle Deschamps has

deux cabinets, l'un de toilette, l'autre de lieux à l'anglaise, le tout orné de glaces,

and though we accept the mirrors in the *cabinet de toilette* we wonder at their introduction into the *anglaise*. A Paris advertisement of 1759 for an apartment to let mentions *Cabinets et Commodités à l'anglaise en aile*, that is, in a separate wing; we swallow the insult in the allocation of the capital letters. Another of 1790 offers *un grand cabinet de toilette et anglaise*. At least these English Places were not ill-furnished; one is fitted up in polished oak, with a marble pan and gilt fittings. Fournier, the historian of inventions, describes them all as *inodores* (not smelly), but ruffles our patriotism afresh by agreeing with Blondel that this English invention is not to be found in England until a much later date. Such innuendoes may have been no more than a balm to Gallic pride, to ease the humiliating acceptance of this sanitary debt. But in the eighteenth century in England the water-closet is admittedly something of a novelty; even in the great houses they are rather rare and rude. At Windsor there is fitted up for Queen Anne 'a little place of Easement of marble with sluices of water to wash all down'. Aubrey in 1718 sees at Sir Francis Carew's house at Beddington in Surrey

a pretty machine to cleanse an House of Office, viz., by a small stream of water no bigger than one's finger, which ran into an engine made like a bit of a fire-shovel, which hung upon its centre of gravity, so that when it was full a considerable quantity of water fell down with some force.

This may have been a self-operating intermittent flushing device, not one to be worked by the user.

William Kent, in his designs for the great house of Holkham in Norfolk, where the kitchen and dining-room are 200 feet apart, gives the Earl of Leicester only a windowless 'two-holer' in an odd corner of the hall. In *The London Art of Building* of the same year (1734) a schedule of plumber's works says nothing of water-closets but

that convenient Cisterns be well placed, plentifully to furnish every Office with Water; and that proper Machines be made to raise the same therein.

Campbell's *Compendious View of All Trades* in 1747 also mentions water for the Office Houses (can no two persons use the same expression?) but says not a word of closets, traps or soil-pipes. At Woburn in 1748 the Duke of Bedford instals a drainage system, complete with four water-closets of which, it is proudly noted, 'at least one is within the house'. In Isaac Ware's *A Complete Body of Architecture* of 1756 a Plan of Sewers and Drains shows external 'Bog Houses' to a very large mansion, but none indoors; it shows a public sewer under the street as well as the usual cesspits of the period, but no 'stink traps' are mentioned. In the vastness of Kedleston, the architect James Paine puts but one single indoor water-closet. Osterley House has one contrived in a niche in one of the rooms, with a door so close to the seat as to hide it only when *not* in use; the drain is brought into the house to connect with its unventilated soil-pipe. As late as 1760 Horace Walpole writes of Aelia Laelia Chudley's house:

But of all curiosities, are the *conveniences* in every bedchamber: great mahogany projections . . . with the holes, with brass handles, and cocks, &c.—I could not help saying, it was the *loosest* family I ever saw!

It is doubtful if these had piped water or drainage, but in any case it is noteworthy that Walpole should rate them as curiosities.

The water-closet of the eighteenth century might be of lead, or even hewn from solid marble, with a metal pan or plunger mechanism, but towards the end of the century the upper bowl might be of glazed pottery. William Hawkins of Fleet Street was then selling some pretty ones, with dainty pictures in willow-pattern style for any who cared to peer closely. In 1824 Hawkins was to advertise a newly-invented self-acting *portable* water-closet, but no details survive, though a tiny bowl and copper pan made by Hawkins exist that may have belonged to one of these.

In the pan closet, a hinged metal pan, when level, kept a few inches of water at the bottom of an upper bowl. When a handle was pulled the pan would swing down, and the contents were supposed to be tipped into a cast-iron or lead receiver and to pass into a trap below. The 'receiver' might have been better termed a 'retainer'. Banner's

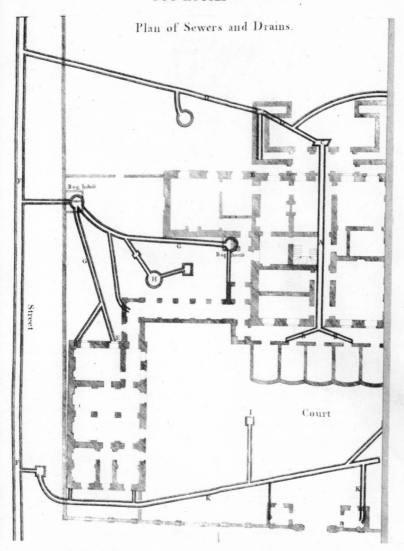

Plan of Sewers and Drains.

Plan of Sewers and Drains by Isaac Ware, 1756

Pan Closet

Patent Drain Trap was an equally objectionable variation. Even with the later additions of a flushing rim and the venting of the container (but often with matters made worse by connecting the vent to the soil-pipe) these horrid devices might need to be 'taken up to be sweetened'. They had an undeserved run of over a century; Stevens Hellyer, the great preacher of proper plumbing, says that hundreds were still being made in 1891, and ends his condemnation with a piece of prose worthy of a nobler ending:

> The light of a candle does not die down all at once. Often in its last flickering moments it extends its flame with so much vigour that a stranger to its ways may be pardoned for thinking that it had recovered its lost energy, and was coming back to life and light again. And so it is with the pan closet.

In the plunger or plug closet, a plug from above was supposed to close the outlet, until a handle was lifted, but this did not remain water-tight, and like the pan closet it had too many foul crannies. Noisome in themselves, these things were connected to unventilated soil-pipes, and so fed sewer gases into the house. Though their defects must have been all too evident, not a single patent for a water-closet of any kind was taken out in the first 158 years of British patents (1617–1775).

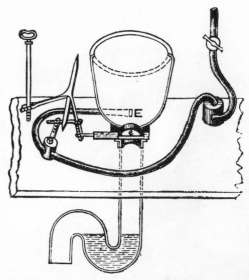

Cummings' Valve closet, 1775

The first such patent, 179 years after 'Ajax', was taken out in 1775 by Alexander Cummings, a watchmaker of Bond Street. In this are found all the elements of the modern valve closet (so called from the valve that closes the outlet of the basin, not from the flushing-water supply-valve): it has the overhead supply cistern, the valve interconnected with the flush and with the pull-up handle, and the syphon-trap. The water is brought into the basin very low down, and is kept in the basin by what Cummings calls 'the slider'.

The advantages of the said water closet depend upon the shape of the pan or bason, the manner of admitting water into it, and on having the stink-trap so constructed that its contents shall, or may, be emptied every time the closet is used.

Only the valve was unreliable; the words 'shall, or may' are significant. With Bramah's improvements to Cummings' valve, a design resulted that was to serve for well over a century.

Joseph Bramah was a cabinet-maker, but one of his jobs was to fit up water-closets, and he greatly improved on Cummings' with a valve that seated itself with a cranking motion instead of loosely sliding. Bramah's patent is dated 1778. By 1797 he had, he said, made about 6,000 closets, and the firm went on making them until about 1890. For all that time they were the accepted pattern, and as

107

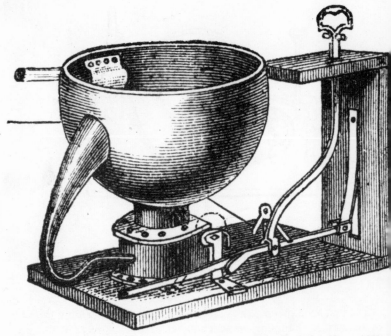

Bramah's Valve Closet, 1778

improved by Hellyer and others, valve closets essentially 'Bramah' are still found in use. The Royal Institute of British Architects had one in their former premises in Conduit Street, that incidentally had the flap of the valve whimsically made of mirror glass. There may even be a few actual 'Bramahs' still working.

With inventions it often pays to get in second, not first. Bramah seems to have taken much of the credit, and all of the business, that might have gone to Cummings, whose patent has all the elements of Bramah's except the hinged valve.

By 1814 a Mr. Phair, writing on the allied subjects of water-closets, chimneys and bell-hanging, can say that water-closets are now in general use and are fitted up in the neatest manner with excellent workmanship. But he has to admit that he knows of families who have to tolerate the same effluvia as if an open churchyard were contained within the walls of the house. He wonders that any gentleman could imagine that air enclosed in a pipe, shut at both ends, with no means of escape, should not be forced through the water valve into the closet, 'of course by mistake', and so discharge stagnated air into

108

WORK FOR THE PLUMBER
Rowlandson, c. 1790

the closet every time it is used, with a 'gulering' noise. A. J. Lamb, who quotes the above, adds an account that shows the risks run by plumbers:

Two workmen had a narrow escape for their lives, for, upon opening the flags, one of them, bowing down to examine the shaft, was suddenly surrounded by flame from a lighted candle in his hand. There was an explosion which split a water bucket, stunned the men, and shut the door, which was half open, with a great noise.

IX

Bason, Bidet and Pot

The French Taste — The Cabinet Makers — Bedroom and Dressing Room — Innocent Trifles of Youth — Evolution of the Wash-basin — A Dainty Tripod—A Perfunctory Dabble—The Corner Wash-basin — Harlequin Furniture — The Dressing Stand — The Bidet Shaving Table — Une Belle Toilette — The Accidental Occasions of the Night — The Night Table — The Bidet — Commode An Ambiguous Word — The Demoiselle — No Shaving With the Zograscope — Pot Cupboards — The History of the Jerry — The Orignal — No Need To Hide It — Earthenware, Lead, Pewter, Tin, Silver, Glass or Gilt? — English Pottery — Crude Early Glazes — White Ware — Printing on China — True Porcelain — Meissen, St. Cloud, Sèvres — The Tyrant and Benjamin Franklin — Jerries For Hire — Jerry Jokes — Salacious Verses — Carriage Pots — The Jerry As An Offensive Weapon — Historical Railway Relics — Chamber Music — An Untimely Entertainment

I N matters of taste, domestic habit and furniture design, the first half of the eighteenth century in England is quite different from the second. In the first half the rougher ways of the seventeenth century persist. One washes under the pump; it is the period of coarse humour, of Hogarth and gin. After 1750 the 'French taste' sweeps society, which comes to prefer wit, Fragonard and wine. The fashionable Englishman becomes more slender and elegant, and so does his furniture, made by Chippendale, Hepplewhite, Shearer or Sheraton, or at least according to their pattern books. He need not be a nobleman; Hepplewhite's book carries no noble dedication but is addressed to 'the residents of London'.

The Age of Elegance justifies its name when we turn from the Bog House to the bedroom and the dressing-room. 'The dressing-room', says Sheraton, 'exhibits the toilet table and commode, with all the little affairs requisite to dress, as bason-stands, stools, glasses, and boxes with all the innocent trifles of youth.' Here we may trace the evolution of the wash-basin from the mediaeval *lavabo*. About 1740, this reappears as a dainty tripod with a tiny china bowl fitting in to a round recess or hole. Below is a shelf for a tiny jug of water, perhaps scented toilet-water. In between there may be another shelf, or tiny drawers for the innocent trifles. A later variant is a tripod with a round case for the bowl, its lid opening to reveal a round mirror. Until marble or porcelain are thought of (in France) as the practical materials for a waterproof top, the polish must sometimes suffer from a spill. But there would be no hearty splashing and spluttering he-man's wash in a little basin of this sort, made for no more than a perfunctory dabble.

About 1770, the wash-stand is moved into a corner. The top becomes a quadrant, with an upstand, is enlarged to take a bigger basin, and has extra holes for round soap-dishes. Sheraton illustrates such pieces.

'Lave-mains', pewter, on wrought iron stand, sixteenth century

Bason Stand by Chippendale, 1754

'Lavabo Vulgaire', 1855

Night Bason Stand by Sheraton, 1803

113

Shaving Table by Hepplewhite, 1787

Soon the famous London cabinet-makers are devising Dressing Stands or Shaving Tables with additional gadgets, and they delight in 'harlequin' or transformable furniture (Harlequin's act in the pantomime being magic transformation) with ingenious sliding and folding parts, exquisitely made, whereby a complex piece closes up into a seemingly simple cabinet. Shearer's Lady's Dressing Stand of 1788 is described in his catalogue:

Two drawers and two sham drawers in front of a square bidet, supported by two drop feet; a glass-frame hinged to a sliding piece, and four cups; a flap to cover the basin hinged to the back of the drawer; a cistern behind to receive the water from the basin drawer; a sweep bidet.

Hepplewhite's *Cabinet Maker's and Upholsterer's Guide* of the same year shows a similar piece.

114

In Hepplewhite's Bidet Shaving Table the mirror rises on a ratchet to any desired height. There is a shallow bidet below that slides forward for use: when the bidet is closed its legs, being triangular in section, meet with the main triangular legs to form square-sectioned legs, and so hide the secret. Similar toilet furniture comes from France; in 1771 the Duchess of Bedford has in her London bedroom a *belle toilette de bois rose plaqué à fleurs de bois violett garnye de porcelaine*, brought from Paris.

Sheraton advises that the 'lodging room' (bedroom) should include such pieces as are necessary for 'the accidental occasions of the night'. Hepplewhite, too, takes care of these occasions, and his Night Table has the same device to hide its stool pot that he used to hide the bidet. The main cupboard above this might hold one or two chamber-pots; it has a 'tambour front', a flexible sliding door made of wooden slats glued to strong canvas, to turn the corner when opened. Even these delicate devices can be found still working smoothly after 170 years' use.

The bidet should be briefly documented here. It is unknown in the seventeenth century, being first mentioned in 1710 when the Marquis d'Argenson was charmed to be granted audience by Mme. de Prie whilst she sat. It is advertised in Paris from 1739, but one early dealer failed to understand its characteristic shape and offered it as 'a porcelain violin-case with four legs'. In 1750 *bidets à seringue* appear, and in 1762 a portable bidet of metal, with removable metal legs carried in the bowl, for the use of officers on campaign. It would clearly be safe to remain seated during a bombardment, for this model is guaranteed shock-proof (*à épreuve des plus fortes sécousses*).

Boucher (who else?) makes one the setting for an original portrait. The Pompadour had two: one cased in rosewood, with floral inlay, gilt bronze legs and fittings, and a bowl of tin (not to be confused with tinplate); the other of walnut, with a cover and back-rest of red morocco with gilt nails, two crystal bottles concealed in the back, and a bowl of porcelain.

Kept at first in the dressing-closet, the bidet moved into the bedroom when secreted in harlequin furniture; a Night Table concealing a bidet appears in Paris in 1783 and may just antedate Hepplewhite's. Thereafter it was hidden away. To the English the bidet has always carried a certain aura of Continental impropriety, and has never quite been accepted. It is found in bathroom designs of the naughty nineties, but then only in the most palatial, made for a sophisticated and well-travelled few, and there it usually has its discreet cupboard. Today in the English middle-class home there cannot be one per thousand bathrooms. The *bidification* of three-star English hotels

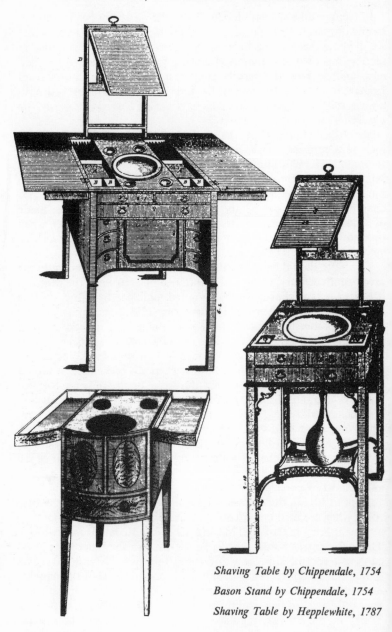

Shaving Table by Chippendale, 1754
Bason Stand by Chippendale, 1754
Shaving Table by Hepplewhite, 1787

Night Table by Hepplewhite, 1787
Shaving Table by Hepplewhite, 1787
Dressing Chest by Sheraton, 1803

117

must compare badly with that of fifth-rate French, fourth-rate Italian or third-rate Spanish. The Ritz Carlton Hotel in New York, having daringly installed bidets some decades ago, was immediately forced by self-appointed guardians of public morals to take them out again. But perhaps the cheap foreign tours of today will in time lead to the discovery by the English (avoiding the word 'British', for Scotland will surely stand firm, though Barrhead make them) and the acceptance of this *meuble utile et discret*.

Returning to the eighteenth-century Night Table: this is known also as a 'commode', an ambiguous word that needs explanation. Chippendale and his school call almost any decorative piece with drawers a commode, as do many museums still, so that some puzzled visitors look in vain for the pot. The bedside table with a pot in a cupboard or drawer, or built in, is at first called a Night Commode; then the general term comes to connote only this particular type. The language of the toilet is indeed an etymologist's nightmare: *chamber* comes by way of *chamber pot* to mean the pot itself; the adjective *privy* (private) comes by way of *privy chamber* to mean the chamber or room itself. *Closet* (small room) comes by way of *water-closet* to mean the apparatus, not the room. *Lavatory* (washing place) comes to mean the water-closet, and to some dainty-minded manufacturers even means the apparatus. *Apparatus* is used here only for want of one accurate word for it. Luckily for confused foreigners, *W.C.* is one of England's three great contributions to universal speech. (The others are *bar* and *sport*.)

French Night Tables

The commode, then, let it be. It includes in its many forms an affair of three steps, each with a countersunk piece of carpet, like the start of an ordinary staircase, used to ascend into the high feather bed of the period, but hiding a sliding stool pot. Later, the commode loses its legs and becomes an ordinary-looking bedside cupboard. There is a small and very portable octagonal or round form that dates well back and long survives; some amateur antique-buyers would be surprised to know what their little coffee table or sewing box originally was.

A rare and rather absurd piece of toilet furniture, known in France as a *demoiselle*, was a sort of robot substitute for a chambermaid. In some forms it was basically a dummy figure for dressmaking, but it could act also as a wig-stand, and given adjustable arms could hold a mirror and a basin in convenient positions; it might include a table for oddments of the toilet. In its near-human form it must have been an unnerving companion overnight in a half-dark bedroom.

The so-called Shaving Stand of the eighteenth century, illustrated in textbooks and seen in antique shops, a turned wooden upright on a round base, carrying a mirror adjustable for height and tilt, and a large hinged convex lens, is not a shaving stand and has nothing to do with the toilet. It is a 'zograscope' for examining engravings, of which it gives an image enlarged but reversed, hence the reversed lettering and topography often seen in eighteenth-century prints.

Hepplewhite shows some jolly little Pot Cupboards in marquetry. These were for the bedroom, but the dining-room had its pot cupboard in one end of the sideboard, or shelves for jerry-pots behind the window-shutter recesses, for the use of gentlemen who drank on after dinner. This useful receptacle the 'jerry' deserves a little history of its own.

The father of the jerry is the mediaeval *orignal*. This was a vase, usually of glass but sometimes of glazed earthenware and very occasionally of metal, with a narrow neck and a wide funnel mouth. It had two advantages. Night-shirts or pyjamas were not yet worn, and the *orignal* could be used without emerging on a cold night from the deep feather bed. If of glass, it was ideal for 'uroscopy', by which art a skilled physician could diagnose all manner of ailments by a mere visual inspection. (Hospitals today use a similar device, but admit that uroscopy has some limitations. Solo airmen have found milk bottles useful.) The Duc de Berry in 1416 had an *orignal* of

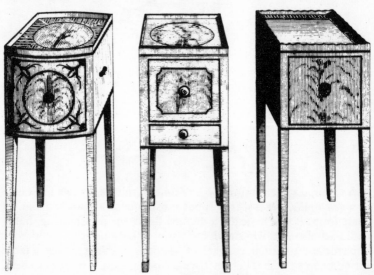

Pot Cupboards by Hepplewhite, 1787

120

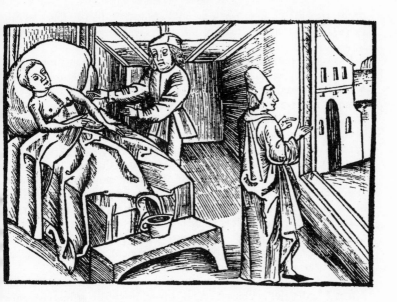

The Mediaeval Jerry

121

decorative glass, hung on four gold chains. Viollet-le-Duc, writing in 1871, says that the form of the mediaeval *orignal* is 'the same as nowadays'.

From the fourteenth century the jerry proper appears; it might be of tin, pewter, copper, silver or gold. Several illustrations show it boldly displayed on a stool at the foot of the bed; it was not yet thought necessary to hide it. Earthenware jerries are recorded from 1418. An inventory of the furnishings of a *château* made in 1471 lists twelve *pots-de-chambre* that might be hard to clean but would be hard to spill, being made of lead. Among the collection of pots stored away by the dispossessed London Museum, the first that can be identified for certain as jerries are of pewter, now rather battered. Older earthenware pots, dating back to mediaeval or even Roman times, may well have been used without being expressly made for this purpose. Among countless such pots in the august shelter of Lancaster House, the jerries mingle inseparably with the cooking-pots even to the expert eye, the forms being identical. The single handle is no clue; the jerry often had two handles.

In 1615 there were twenty-two tin jerries at the *Château de Turenne*. James I had a richly-chased one of silver. In 1653 Cardinal Mazarin had a glass one covered with velour, decorated with a gold band and with silk and gold tasselled cords. Louis XIV had two of silver, carrying his arms, and two similar 'spares', but he went to war with a special one of *argent vermeil doré*.

The development of a native English pottery began in North Staffordshire, where a growing community of peasant potters was established about the mid-seventeenth century. Their first earthenware jerries were clumsily fashioned for use rather than for show, and if glazed at all, only with coarse yellow or green lead glazes, often applied only inside. In the mid-eighteenth century the 'Potteries' found the art of making ware white throughout, by mixing selected white-burning clays with finely-ground flint (silica), and the china jerry could then already meet the specification for the material of sanitary appliances to be laid down 200 years later by the British Standards Institution, whereby it should be 'durable, impervious, non-corrodible, and have a smooth surface which can be easily cleaned'. This fine white ware made a good base for coloured glazes, and then (with the invention of printing on china in 1754 by Mr. Sadler of Liverpool) for transfer-printed patterns. The decorated jerry, like the closet-pan and the wash-basin, became a thing of beauty. Armigerent families ordered fine crested ones. True porcelain, introduced into Europe from China in the sixteenth century but not successfully imitated until about 1710 at Meissen, first appears in this field in 1746 when *pots-de-chambre ronds du Japon* are made at St. Cloud.

At an exhibition of Sèvres porcelain at Versailles in 1776, a medallion portrait of the American envoy Benjamin Franklin, with the republican legend *Eripuit coelo fulmen, sceptrumque tyrannis*, was being tactlessly sold under the eyes of Louis XVI. Ignoring the insult, the 'tyrant' quietly ordered from Sèvres a chamber-pot containing both portrait and legend, and thereupon relieved his feelings.

Parson Woodforde, describing the great frost of 1785, notes in his diary that the chamber pots froze under the beds.

In the eighteenth century jerries could even be hired: the municipality of Rennes, for a ball at the Town Hall, hired fifty at 3 sols apiece; it must have been a fairly orderly party, for they had to pay for only two that were broken.

By about 1800 some were very elaborate, with applied modelling in full relief: flowers outside, or perhaps a realistic frog inside. Jokes fall into a few main classes, of which one depends on the breach of a taboo, so that the jerry became a sure source of naughty fun, and could be designed accordingly. One could have a portrait of Napoleon on the target area, or a large eye with the lines

> Use me well, and keep me clean,
> And I'll not tell what I have seen

—a jest so popular that it is still to be found, in miniature form suitable for the mantelpiece, in seaside postcard shops. A splendid example in Sunderland lustre ware, now in the Leicester City Museum, has everything: the frog inside, the lines quoted above, and further 'salacious verses' (as the catalogue calls them) that are furtively copied down by a surprising number of respectable-looking visitors.

'Carriage folk' had 'carriage-pots' under the seats of their vehicles; often the seat was pierced so that it was only necessary to lift the cushion.

A separate chapter could be written on the history of the jerry as an offensive weapon. It appears in this role at an early date, for in 1418 one Baudet was banished from the city of Paris for having broken one on a lady's head:

Le dit Baudet, courroucie de ce, print le pot, a quoy il se aisoit aucunes fois de nuit . . . et le lui gecta at l'en assena par la teste ou par les epaules, telement que le pot cassa.

The painter Greuze (that gentle painter, of all painters), was similarly assaulted by his wife, and that normally brave warrior the

123

Prince de Condé admitted his terror of this shattering missile when street-fighting came to *la guerre de pots de chambre.*

Many of the later nineteenth-century jerries are fine examples of the potter's art. A priceless collection guarded by the Curator of Historical Relics to the British Transport Commission includes:

Great Central Railway Co. Blue crest and linings. (Spode.)

L.N.E.R. Marine. Light blue crest and linings. (Minton.)

Great Northern Hotel, Kings Cross. Green arabesques, crest on bottom. (Doulton.)

Manchester, Sheffield & Lincolnshire Railway Co. Blue crest and linings. (Spode.)

The Bath Hotel, Felixstowe. Turquoise lining round rim: roses in red and green. (Doulton.)

Norfolk Broads, L.N.E.R. Scenes of The Broads, windmills &c., in blue, on the outside and inside the rim. (Doulton.)

A novelty of the late nineteenth century is the jerry with a concealed musical box, that gives a recital of appropriate chamber music when it is lifted, and cannot be turned off by the embarrassed guest. But not truly a novelty: in 1820 Prince Metternich, staying at the Palace of Furstenberg, was awakened at midnight by music of the flute near his bed. His Night Table was giving this untimely entertainment. He found and pressed a button, and the music stopped—more or less, for once an hour it tried to resume, and made disturbing little noises. When he complained in the morning, the valet remarked that the Princess' own Night Table even played *trumpet* music—

les jeunes mariés se fatiguent facilement, et cela les fait dormir.

X

Baths and Buggs

IT was the chilly impact of marble, especially as the bath habit spread to households where servants were not kept by the dozen and sodden cushions and linen were unwelcome, as well as its weight and cost, that led to the use of metal. A copper bath appears among the effects of the Archbishop of Bordeaux in 1680. This was all of metal, except its wooden base and cover, but it was more usual to line a wooden bath with copper or lead. We read of a copper bath of 1759 complete with a water-tank and heating furnace. Perhaps the first advertisement offering a house with a bathroom is one in a Paris paper of 1765. Soon after this, such advertisements are frequent, and there are also many *salles de bains portatives* that would be erected in a courtyard or garden.

Copper was for a long time the favourite material; it was malleable and rustless, but it was expensive. Tin was sometimes used. Sheet iron rusted, and the casting of an iron bath was well beyond the technical skill of the time. In 1770 a M. Clement invented a varnish that could be applied to a sheet-iron bath, and more or less withstood hot water, and thereafter, according to Havard, marble baths were found only *chez certaines beautés aimables et célèbres*.

In 1778 when Mlle Devise of the *Opéra* made a 'moonlight flit', the only item she left perforce in her apartment was a copper bath, upholstered in wickerwork, with a copper cylinder. Such cylinders were heaters. They were hardly equal to the capacity of a full-sized bath, and M. Level of Paris instead of scaling up the heater, scaled down the bath to a small one made like a chair. His heater ran on spirit, an expensive fuel, and he experimented with coal and coke, but ran into trouble with the fumes. No attempt seems to have been made to pipe the hot water from a heater in another room, probably because the temperature of the bath-water was already marginal.

We find baths of this period made as pieces of fine furniture: round ones like chairs or long ones like sofas, with padded backs and all the trappings of the *ébéniste* or cabinet-maker. Designed so evidently to impress, they may well have featured in the *levées* held in the bed-

'Bain en Forme de Sopha'

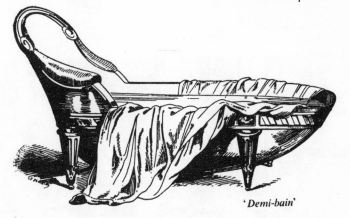

'Demi-bain'

rooms of the great. Erotic prints show a type of *chaise longue* made to open up, revealing a shallow bath within. Casanova's bath in his Paris apartment was portable—it could be put at his bedside—but it had room for two. The *demi-bain* from Havard, though of metal, is draped with loose fabric as the marble baths were; if this fabric was left for long as it is shown in the picture, capillary attraction and syphonic action must surely have combined to empty the bath on to the carpet.

127

'Rejuvenation', 1600

Engraving from a mural painting by Raphael for Cardinal Bibiena's bathroom in the Vatican

128

Le Beau, 1773

Throughout history, Eros flits in and out of the bathroom—even in the Vatican

The *baignoire en berceau* or cradle-bath is a frail wooden cockle-shell with a metal lining, suspended to swing between two *rocaille* supports; a sort of nursery version of the Roman *pensiles balneae*; it is all too easy to empty, and may have given rise to the expression 'throwing out the baby with the bath water'.

Queen Elizabeth II, and other distinguished guests of the President of the French Republic, have lately enjoyed the splendours of a restored bathroom in the *Palais de l'Elysée* said to have been fitted up for Napoleon (who took a very hot bath daily—indeed, one historian has suggested that the course of history might have run differently had he not thereby sapped his energies. Wellington took a cold bath daily; perhaps Waterloo was won in the bathroom). Were the bath removed, this apartment might serve well as a small ballroom. Mirrors enriched with painted decoration reflect a magnificent chandelier. The bather, emerging from the bath, may pad across the splendid carpet to dry by the great white marble fireplace.

The Empire style bathroom made for Eugène de Beauharnais by Boffrand (still existing in the *Rue de Lille* in Paris) has walls wholly faced in mirror glass, reflecting an infinite forest of columns among which an infinite army of bathers carries out perfectly synchronised bath-drill. Another impressive bathroom still to be seen in Paris is in the former *hôtel* in the *Champs-Elysées* now housing the Travellers' Club. Made for the famous cocotte La Païva during the Second Empire, in Moorish style, it uses onyx, silver-plate for the bath, turquoises to encrust the taps; all under a Moorish ceiling with a

La Païva's Bathroom, Champs-Elysées, Paris

Engraving by Girard from 'A Day in the Life of a Courtesan'

cornice of glass stalactites. Girard depicts a bath of clear crystal, of which no example survives, whose lovely occupant can never be hidden. At the Elector's Palace at Schwetzingen in the Grand Duchy of Baden, rebuilt in the mid-eighteenth century in the manner of Versailles, the best bathroom, still to be seen, has a floor of pink and black marble, and walls of white stucco with nymphs in relief in panels bordered with amethysts. Imitation draperies in stucco surround the bath, which is sunk in the floor and reached by three steps. The water is delivered through the mouths of eight serpents that coil round the margin of the bath.

132

The Royal apartments in the Pitti Palace at Florence have a neo-classic bathroom so vast and imposing that any attempt to visualise its effect when the bath is occupied by the average human figure produces a rather ridiculous result. In a room of such proportions, man has less chance than ever of seeming a hero to his valet. Even a cake of soap would look silly here.

Ranson: Design for a Bathroom: Period Louis XVI

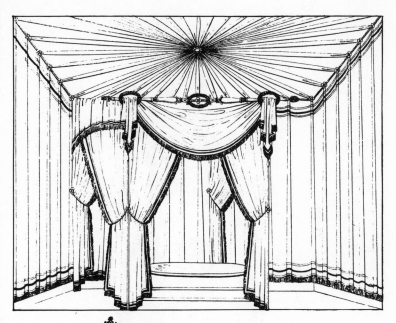

French Bathroom, c. 1800

Wash Stand by Percier, c. 1800

135

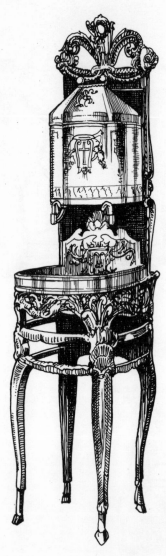

*'Fontaine Lavabo': wood, with reservoir and basin of beaten and engraved copper:
French, late eighteenth century*

Le Plaisir de l'Été (1741)

The contrast between the noble and the bourgeois bath survives
the Revolution. Until well into the nineteenth century, Paris water-
sellers still carried bath-tubs in their carts, complete with hot water,
and brought them up to their customers' apartments. In 1838 there
were 1,013 of these *baignoires à domicile* duly recorded and licensed,
and only 2,224 fixed baths. Such luxuries were even more rare in
Britain. A private lunatic asylum opened in Norwich in 1765 had hot
and cold baths to quell unruly patients, but the sane seldom installed
them. *The Edinburgh Advertiser* in 1770 announced the opening of
'A Neat Bath for Gentlemen and Ladies', but all we know of this is
that every person was served with fresh water. In 1789 Shardeloes
House in Buckinghamshire had a new 'iern-bound bathing tub' costing 4 guineas, but in the same year Howard, the prison-reformer,
noted that although Guy's Hospital had excellent baths, and water-
closets flushed automatically by the opening of the doors, St.
George's Hospital had only one bath and never used it; the London
Hospital sometimes used its one bath, but this was kept in a dirty
cellar, and served 126 patients. A Dr. Lucas, whose habits were

cleaner than his spelling, proposed—doubtless in vain—that bath
should be provided in 'bridewels, goals and other places of confine
ment'.

A manual of etiquette dated 1782 advises wiping the face ever
morning with a white linen, but warns that it is not so good to was
it in water, for that makes the face too sensitive to cold and sunburr
A doctor writing in 1801 remarks that 'most men resident in Londo
and many ladies though accustomed to wash their hands and face
daily, neglect washing their bodies from year to year'—at least th
ladies seem to have been a shade nicer than the men. Charles Howar
11th Duke of Norfolk, known as 'Jockey', did not go quite unwashe
as he would have preferred, for according to Osbert Sitwell an
Margaret Barton

the Duke possessed a capacity for liquor no less remarkable than h
digestion; he would drink everyone else under the table and then mo
on to finish the evening elsewhere. Moreover, until suddenly he becam
speechless and immovable in his chair, he never betrayed any symptom c
intoxication. At a sign from him as he was about to lose consciousness,
servant would ring the bell three times; when four footmen would imm
diately answer it, bearing a kind of stretcher. In absolute silence, and wit
a dexterity that betrayed long practice, they lifted him on to it and, with
gentle swinging motion, removed his enormous bulk from the room. If h
became quite insensible and unable to resist, they would sometimes tak
it upon themselves to remove his clothes and scrub his body with soap an
water; a highly necessary rite, since the Duke hated the sight of water an
remained determined to ignore its existence.

In 1812 the Common Council turned down a request from th
Lord Mayor of London for a mere shower-bath in the Mansio
House 'inasmuch as the want thereof has never been complained of'
if he wanted one, he might provide a temporary one at his ow
expense. Twenty years later, however, they had agreed not only t
a bath, but to some sort of hot water supply, perhaps because of th
examples of the Duke of Wellington (who as has been mentione
took a cold bath daily), and of Lord John Russell, later Prim
Minister, who had designed for himself a great mahogany bat
lined with sheet lead, that weighed a ton.

An advanced thinker of 1829 did not disparage the 'hydromania
fever' that sent the family to the sea in August, but even held tha
bathing ought to be a daily instead of a yearly practice. Saner opinor
recognised that frequent bathing must increase rheumatic fever an
lung complaints. Cobbett, known to be a dangerous radical, in hi
Advice to a Lover (1829) held that cleanliness is

a capital ingredient; for there never yet was, and there never will be, love of any long duration, sincere and ardent love, in any man towards a '*filthy mate*'; I mean any man in *England*, or in those parts of America where the people are descended from the English.

So he advised the lover to look behind the lady's ears. (Americans who had not the advantage of English descent presumably continued to love their filthy mates.) Of the other school of thought, one of the Georgian Royal Dukes remarked that it was sweat, damn it, that kept a man clean.

At Queen Victoria's accession in 1837 there was no bath in Buckingham Palace, but Parliament voted her £5,000 a year for her 'toilette', and although this included her clothes, plans were drawn up for conveying hot water to Her Majesty's portable bath in her bedroom. Meanwhile she had inherited at least one bath: George IV's bath in the Royal Pavilion at Brighton; but this, measuring 16 ft. by 10 ft. by 6 ft. deep, was not a large bathtub but a small swimming bath, supplied with sea water by 'a succession of pipes and other machinery'. One does not somehow see the Queen using it, and she did in fact have it demolished; the white marble from this bathroom was sawn up to make mantelpieces for Buckingham Palace. Such trend as there was towards cleanliness owed little to Royal example. No objection seems to have been raised at the Palace when Messrs. Tiffin & Son put up their advertisement:

TIFFIN & SON,
BUG-DESTROYERS TO HER MAJESTY.

—but perhaps this *was* in a way a good example. The senior partner in this house kindly obliged Mayhew, the tireless cataloguer of the details of London life, with a statement:

We can trace our business back as far as 1695, when one of our ancestors first turned his attention to the destruction of bugs.

. . . I knew a case of a bug who used to come every night about 30 or 40 feet—it was an immense large room—from a corner of the room to visit an old lady. There was only one bug, and he'd been there for a long time. I was sent for to find him out. It took me a long time to catch him. In that instance I had to examine every part of the room, and when I got him I gave him an extra nip to serve him out.

. . . I work for the upper classes only; that is, for carriage company and such-like approaching it, you know. I have noblemen's names, the first in England, on my books.

139

... I was once at work on the Princess Charlotte's own bedstead. I was in the room, and she asked me if I had found anything, and I told her no; but just at that minute I *did* happen to catch one, and upon that she sprang up on the bed, and put her hand on my shoulder, to look at it. She had been tormented by the creature, because I was ordered to come directly, and that was the only one I found. When the Princess saw it, she said, 'Oh, the nasty thing! That's what tormented me last night; don't let him escape.' I think he looked all the better for having tasted royal blood.

Messrs. Tiffin were not the first claimants to the honours of Royal Disinfestation: in 1775 Andrew Cooke of Holborn Hill, who boasted of having 'cured 16,000 beds with great applause' and of having worked at the Palace, resented that a rival should advertise himself as 'Bug-destroyer to His Majesty'. Earlier still, in 1740, Mary Southall, though she could not claim Royal patronage, could claim the approbation of the Royal Society:

MARY SOUTHALL

Successor to John Southall, *the first and only person that ever found out the nature of* BUGGS, *Author of the Treatise of those nauseous venomous Insects, published with the Approbation (and for which he had the honour to receive the unanimous Thanks) of the Royal Society,*

GIVES NOTICE,

THAT since his decease she hath followed the same business, and lives at the house of Mrs. Mary Roundhall, in Bearlane, Christ Church Parish, Southwark. Such quality and gentry as are troubled with buggs, and are desirous to be kept free from those vermin, may know, on sending their commands to her lodgings aforesaid, when she will agree with them on easy terms, and at the first sight will justly tell them which of their beds are infested, &c., and which are free, and what is the expense of clearing the infested ones, never putting any one to more expense than necessary.

Persons who cannot afford to pay her price, and is willing to destroy them themselves, may by sending notice to her place of abode aforesaid, be furnish'd with the NON PAREIL LIQUOR, &c. &c.

If we are to believe Southall's *Treatise of Bugs*, the valuable addition to zoology quoted above, the horrid story can not be taken much further back, for he affirms that the insect was scarcely known in England before 1670, when it was imported among the timber used to rebuild the city of London after the Great Fire. But Dr. Thomas Muffet in his *Theatrum Insectorum* tells us that Dr. Penny, a previous compiler of that work, has a tale of being sent for in great haste to visit two noble ladies of Mortlake who imagined themselves seized with symptoms of the plague; the cause proved to be 'buggs', and

140

Jack Black, Her Majesty's Ratcatcher

141

this was in 1583. None of these experts seems to have hit upon the interesting method observed in Persia by Freya Stark: a narrow channel filled with water around the walls of the room, which deterred or drowned the creatures.

Queen Victoria had the first railway lavatory compartment, when the first Royal Saloon was built in 1840 by the Great Western Railway. This was for the old broad gauge (7 ft.) and had six wheels. The pedestal for the china wash-basin conceals a bucket for the waste water, and the semi-circular front is tastefully upholstered in quilted watered silk; this pretty piece is preserved by the British Transport Commission. The South Eastern Railway Company's Royal Saloon of 1850 had a 'patent convenience' hidden in a sofa. Not until the 1860's was anything provided for commoner clay, and then only in private saloon compartments. By 1874 the first British Pullman coaches had valve water-closets, and by 1881 on the Midland Railway even third-class passengers were granted relief.

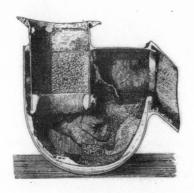

XI

The Cholera Years

Town Sanitation Declines — Leeds in the 1830's — Death Rates
— St. Giles' Rookery — Discoveries Under Westminster Abbey
— The Back-to-Backs — A Cesspit Under the Living Room —
Night Men — The Fleet Ditch Again — Contaminated Wells
— Accident to a Carriage — The Fever — Signs of Bad Weather
— Under Windsor Castle — Wimpole Street and Harley Street
— Mr. Eassie's Cat Dies — The Squire's Darter Dies — Irregular
Water Supply — Iron Water Mains — Thames Water — Dry
Sunday in Bristol — Water Carriers — Cholera — Edwin
Chadwick — Charles Kingsley — The Sanitary Idea — Public
Baths and Wash Houses — Doulton of Lambeth — Plain Sheds
and Cabins — St. Marylebone Baths — George St. Baths — The
First Public Health Act — The Nuisances Removal Act — State
of the Thames — The Defenders of Filth — The Sewer
Commissioners' Survey — Fashionable Squares — Sewer Rats
— No More Night Men — Joseph Bazalgette — London's New
Drainage System

143

IN the bigger towns a network of drains was slowly spreading, but with ignorant or skimped workmanship and bad maintenance some of these in time did as much harm as good. In the first thirty years of the century the population of Greater London almost doubled, to 1½ millions, and in the next twenty years another million were somehow crowded in. With this teeming humanity largely housed by slum landlords, with the *laissez-faire* philosophy current, and with apathy towards hygiene at all social levels, the standard of town sanitation sank well below that of the unpiped countryside. R. H. Mottram in *Early Victorian England* quotes a report on conditions in Leeds in the 1830's:

> 568 streets were taken for examination: 68 were paved: 96 were neither paved, drained nor cleaned; one of them, with 176 families, had not been touched for 15 years. Whole streets were floating with sewage; 200 were crossed with clothes-lines. Over 500 cellars were in occupation. 156 rudimentary schools provided for 7,000 children; the Sunday-schools took in 11,000; 15,000 went altogether untaught. Finally we learn there were 451 public houses, 98 brothels, 2 churches, and 39 meeting-houses. The death rate in the clean streets was 1 in 36; in the dirty streets 1 in 23.

At this time the death-rate per 1,000 under five years old was 240 in the country (Herefordshire) and 480 in the town (Nottingham). Today five times as many children survive their infancy as did in 1848. No townsman could ever be really well. Over half a million were dying every year from preventable causes arising mainly from living conditions. The old London rookery of St. Giles had 95 small houses containing 2,850 people, and was flooded by its own sewage. An outbreak of fever at Westminster led to the discovery of old cesspits and sewers under the Abbey precincts from which 500 cartloads were cleared away. The new jerry-built working-class houses, though they had no cellars to house rats or humans, were built in double rows back-to-back, without ventilation or drainage, their windows cut to a minimum number to avoid the Window Tax. They formed squalid courts, with a pump at one end and a privy at the other to serve perhaps twenty dwellings. Well did Rousseau say that man is, of all animals, least calculated for the social state.

Even in the best houses the cesspit persisted; it might be in the yard, or inside the house; even below a living-room floor. The contents had to be carried away in buckets, through the house if the plan so required, just as in Pepys' time. 'Night Men', successors to the mediaeval *gongfermors*, performed this office, as witness an elegantly-engraved trade card, now in the London Museum:

The Fleet River near St. Pancras in 1825

The Fleet Ditch in 1844, with house drains discharging directly into the water

At first the value of the end-product to farmers made it worth while to remove it from the town, but as London grew the journey became uneconomical. The buckets were then carried (or the 'new invented Machine Carts' were driven) to larger communal cesspits, or to the nearest river, which might still provide the drinking water. The Fleet Ditch as depicted in 1840 looks no more savoury than it had been 500 years earlier when Edward III attempted its cleaning, but only in 1841 was it recognised as a sewer and covered in; it still runs, but innocuously, under Farringdon Street and New Bridge Street.

The Fleet Sewer during the building of St. Pancras Station. 'The Fleet Sewer affair involved the taking up of a main artery of metropolitan drainage, the diversion of a miniature—indeed scarcely a miniature—Styx, whose black and foetid torrent had to be transferred from its bed of half-rotten bricks to an iron tunnel running in an entirely different direction, and that, too, without the spilling of "one drop of Christian" sewage'

If the drinking water came from a well, it might be contaminated by a leaky cesspit nearby. It was common for a full pit to be covered over and forgotten, and a new one dug. At one country mansion attention was drawn to this habit by the sudden sinking of a carriage as it pulled up at the front door. No wonder that in the novels of the day, the convenient way to put a character to death, or at least to bed, was with 'the fever'. Lady Georgiana Russell in her *Recollections* tells us that

in those careless days (1837) . . . bad drains were considered rather a joke. If they smelt, people considered it a sign of bad weather approaching and were rather pleased to have the warning.

In 1844 no less than fifty-three overflowing cesspits were found under Windsor Castle, to explain the sore throats and worse ailments long suffered by the servants. The Prince Consort took matters in hand, and replaced the Hanoverian commodes with water-closets and drains, but with his death, his widow's rule that all must remain exactly as he left it included the sanitation.

Eassie describes a house in Wimpole Street with a brick-barrel drain passing under it, and without a proper fall, where none of the solid sewage or wastes ever reached the sewer in thirty years. At another in Harley Street, where the drain was lower than the sewer, forty barrowloads were taken away. At a third, a plumber used a candle as he raised the flags under the sink, and was blown up. At a fourth, where Eassie had the floor taken up in search of rats,

over 50 nests were laid bare, besides a goodly collection of all kinds of bones, flanked by once shining articles, such as scissors, corkscrews, and one or two silver spoons. A favourite cat, and a good mouser, which I had sent there before I opened up the floors, died after a few days experience of the place; it was said by some from the effects of nocturnal warfare, although it is just as likely that it was from grief, and because he had, unlike Alexander, too many worlds to conquer.

It is of the squire's darter, not of a cottager's, that the poet is speaking in the painful lines from *The Village Wife*:

Fur hoffens we talkt o' my darter es died o' the fever at fall:
An' I thowt 'twur the will o' the Lord, but Miss Annie she said it wur draäins.

The water supply was still operated with regulated irregularity: each house received water for two or three hours on about three days a week, when the tanks would be filled. A writer of 1802 says proudly that water is conveyed three times a week into almost every house by leaden pipes and preserved in cisterns or tubs in such quantities that the inhabitants have a constant and even lavish supply.

When a fire broke out outside 'opening hours', a messenger was sent running to the water company's offices. The dry periods shortened the life of the mains, and led to leakage and contamination. The New River Company was still using wooden mains in 1802, but from 1827 all new mains had to be of iron, and the New River became an iron pipe, shortened to 27 miles. Although an attempt to draw water from the Thames near Vauxhall Bridge had been abandoned in 1805, and the intake moved to a cleaner reach upstream, the Lambeth Water Company had intakes at Battersea and at Charing Cross until 1848.

In Bristol in the 1830's, though they no longer recorded disbursements for 'taking dead cats out of the conduit' there, of 3,000 houses examined, 1,300 had no water. The Puritan Corporation closed all the conduits on Sundays, and it was a punishable offence to draw water on that day. In 1850 there were 80,000 waterless houses in London, and according to Mayhew there were still from 100 to 150 water-carriers at work, charging 1d. a pail and earning 6d. to 1s. a day. Some families spent up to 2s. a week on carted water. Fortunately water was not so popular as a beverage as it is today; even charity children were given small beer.

'Fever' was accepted, but cholera was something very different. Though it had spread slowly westwards from India and had by 1830 reached European Russia, cholera was thought of as an Asiatic disease incapable of attacking a decent Englishman, until it struck London with sensational effect in 1832, and again at intervals until 1866. Society was at last scared into action. The cholera 'bug' had not yet been identified, but it was known that contaminated water is by far the most common means of dissemination, so that a modern water-supply becomes itself a menace: a single contaminated source can endanger the whole community. Cholera may have come because of, not in spite of the new water-closets. The drained houses of the rich could be a greater menace than the primitive but self-contained filth of the slums. As Mr. Punch, cartooned in the character of Hamlet, asked:

Why may not imagination trace the remains of an Alderman till we find them poisoning his Ward?

Reformers such as Edwin Chadwick, who as a Poor Law Commissioner had seen the evils for himself, and the Rev. Charles Kingsley, author of *The Water Babies*, preached 'The Sanitary Idea'. The

HAS

DEATH

(IN A RAGE)

Been invited by the Commissioners of Common
Sewers to take up his abode in Lambeth? or, from
what other villanous cause proceeds the frightful
Mortality by which we are surrounded?

In this Pest-House of the Metropolis, and
disgrace to the Nation, the main, thoroughfares
are still without Common Sewers, although the
Inhabitants have paid exorbitant Rates from time
immemorial!!!

" O Heaven! that such companions thou'dst unfold,
" And put in every honest hand, a whip,
" To lash the rascals naked through the world."

Unless something be speedily done to allay
the growing discontent of the people, retributive
justice in her salutary vengeance will commence
her operations with the *Lamp-Iron* and the
Halter.

SALUS POPULI.

Lambeth, August, 1832.

J. W. PEEL, Printer, 9, New Cut, Lambeth.

Lord Mayor launched a fund to build cheap public baths and wash-houses, and the first of several Public Baths and Wash Houses Acts was passed in 1846, the year in which Sir Henry Doulton established his factory at Lambeth to make the new glazed stoneware pipes that made really sound drains possible. The first public bath-house was built in Glasshouse Street, Smithfield. Baths, it was stated, could now be enjoyed with no expense beyond that of constructing 'plain sheds and cabins'. How plain such sheds and cabins could be is shown in Doré's engraving of a bath-house of about 1870. A more impressive establishment in St. Marylebone had 107 separate baths, vapour baths, showers and two swimming-baths. Adjoining public wash-houses shared the hot water system. 'Labouring classes' could have a cold bath for 1d., a warm shower or vapour bath for 2d., both charges including a clean towel. 'Those of any higher class' could be charged up to three times this rate: bathers who were anxious to pre-serve both caste and cash must have found it difficult to agree their status. Mayhew records that 96,726 baths were taken in the George Street Baths, Euston Square, in 1849, but the number of 'the great unwashed' was and always has been out of all proportion to the number of public baths. Up to 1865 only twenty-five boroughs had provided any. Even in 1908 in London there was only one municipal tub for about 2,000 inhabitants. The Cheyenne Indians, the Hawai-ians, the Baganda of East Africa and the Chiriguano of the South American Chaco were still well ahead of the Londoner in the matter of daily bathing.

At the Berlin Hygiene Exhibition in 1883 a Dr. Lassar showed the 'People's Baths' (*Volksbaeder*)—a corrugated iron shelter partitioned into cubicles, with hot showers at 10 pfennigs a head—about as inviting as a public urinal, but better than nothing at a time when Germany had one public bath per 30,000 inhabitants.

The first Public Health Act dates from the cholera year of 1848, but it gave permission rather than orders for action, and was not properly carried out by the municipalities for at least twenty years. The Nuisances Removal Act of 1849 gave better powers, and nuis-ances were not far to seek. In that year the sewers delivered over nine million cubic feet of sludge into the Thames, and the scour of the tides had long since failed to carry away more than a fraction of it. The Chelsea Water Company still had its intake a few feet from the outfall of the Ranelagh sewer, formerly the West Bourne. 'We are paying the companies collectively', wrote *The Spectator*, '£340,000 per annum for a more or less concentrated solution of native guano.' In 1852 it was made illegal to draw water from the river below Teddington Lock. Over 14,000 died of cholera in London alone in

A Night Refuge in London: Doré, 1877

Doré (1877)

1849; over 10,000 in a fresh outbreak in 1854; over 5,000 in the last in 1866. In the hot summer of 1858 there was talk of transferring Parliament, whose windows on the river front had not been opened for years, to a tolerable distance from the stench of the Thames. *The Quarterly Review* joined battle in fine style:

> Like another Troy, this citadel of filth has stood a ten years' siege: and its sturdy garrison, led by their chieftains in Common Council—the Hectors and Memnons of intramural muck—so far from thinking of surrender, are engaged at this moment in fortifying their defences. . . . The Defenders of Filth, corporate and parochial, have ruled London long enough.

Outside the city was a patchwork of local authorities with little power, but in 1848 the control of London's drainage, hitherto divided among eight different (and indifferent) bodies, was vested in a Board of Commissioners. Within two years they had almost got rid of cesspits in London, but when they came to attack the old sewers they met an appalling task. Mayhew describes their preliminary survey:

153

The deposit has been found to comprise all the ingredients from the breweries, the gas-works, and the several chemical and mineral manufactories; dead dogs, cats, kittens, and rats; offal from slaughter-houses, sometimes even including the entrails of the animals; street-pavement dirt of every variety; vegetable refuse; stable-dung; the refuse of pig-styes; night-soil; ashes; tin kettles and pans (panshreds); broken stoneware, as jars, pitchers, flower-pots &c.; bricks; pieces of wood; rotten mortar and rubbish of different kinds; and even rags.

. . . a sewer from the Westminster Workhouse, which was of all shapes and sizes, was in so wretched a condition that the leveller could scarcely work for the thick scum that covered the glasses of the spirit-level in a few minutes after being wiped . . . a chamber is reached about 30 feet in length, from the roof of which hangings of putrid matter like stalactites descend three feet in length. At the end of this chamber, the sewer passes under the public privies, the ceilings of which can be seen from it. Beyond this it is not possible to go.

. . . On the 12th January we were very nearly losing a whole party by choke damp, the last man being dragged out on his back (through two feet of black foetid deposits) in a state of insensibility.

154

The sewerage of the aristocratic parts of the city of Westminster and of the fashionable squares: Belgrave, Eaton, Grosvenor, Hanover, Berkeley, Cavendish, Bryanston, Manchester and Portman Squares, was no better than elsewhere:

There is so much rottenness and decay that there is no security for the sewers standing from day to day, and to flush them for the removal of their 'most loathsome deposit' might be 'to bring some of them down altogether' . . . throughout the new Paddington district, the neighbourhood of Hyde Park Gardens, and the costly squares and streets adjacent, the sewers abound with the foulest deposit, from which the most disgusting effluvium arises; indeed, amidst the whole of the Westminster District of Sewers the *only* little spot which can be mentioned as being in at all a satisfactory state is the Seven Dials.

Perhaps the better state of the sewers under this notorious slum was due only to the fact that its inhabitants rarely used them. Mayhew rounds off his picture with a chapter on the rats in the sewers, 'as big as good-sized kittens' and 'fighting and squeaking there like a parcel of drunken Irishmen'. The emergence of sewer rats into the house was an accepted and useful indication of the larger flaws in the drains.

Mayhew's interview with a 'cesspool-sewerman' who had taken up this work because he did not like 'the confinement or the close air in the factories', shows that in the 1850's the replacement of cesspools by sewers was well under way, and that the Night Man had nearly had his day:

The houses we clean out, all says it's far the best plan, ours is. 'Never no more night men', they say. You see, sir, our plan's far less trouble to the people in the house, and there's no smell—least, I never found no smell, and it's cheap, too. In time the night men'll disappear, in course they must, there's so many new dodges comes up, always some one of the working classes being ruined. If it ain't steam, it's something else as knocks the bread out of their mouths quite as quick.

Not until the 1870's was the London death-rate to fall decisively: it could hardly be by coincidence that this fall came within five years of the opening of the new drainage system. As Chief Engineer to the Board of Works, Joseph Bazalgette had fought through his great plan, with 83 miles of large intercepting sewers, draining over 100 square miles of buildings, carrying 420 million gallons a day and costing £4,600,000. This system was opened in 1865. Bazalgette also built the Albert Embankment and the Victoria Embankment where his well-earned and little-known monument now stands.

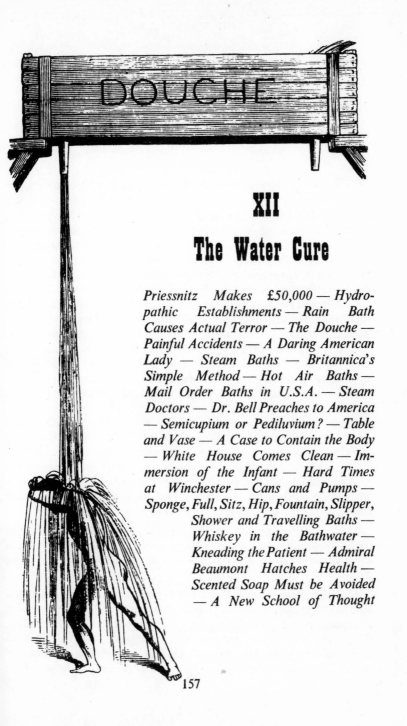

XII

The Water Cure

Priessnitz Makes £50,000 — Hydropathic Establishments — Rain Bath Causes Actual Terror — The Douche — Painful Accidents — A Daring American Lady — Steam Baths — Britannica's Simple Method — Hot Air Baths — Mail Order Baths in U.S.A. — Steam Doctors — Dr. Bell Preaches to America — Semicupium or Pediluvium? — Table and Vase — A Case to Contain the Body — White House Comes Clean — Immersion of the Infant — Hard Times at Winchester — Cans and Pumps — Sponge, Full, Sitz, Hip, Fountain, Slipper, Shower and Travelling Baths — Whiskey in the Bathwater — Kneading the Patient — Admiral Beaumont Hatches Health — Scented Soap Must be Avoided — A New School of Thought

SLOWLY the fear of water began to give way, as more doctors prescribed the various forms of Water Cure, though it was still thought eccentric to bathe for any but medical reasons. The undisputed inventor of the *true* Water Cure was Vincent Priessnitz, a Silesian peasant, who opened his establishment at Grafenberg in 1829. At the age of thirteen he had cured his sprained wrist by a wet bandage, and he experimented further with the healing fluid. At the age of sixteen, being trampled by a horse, he broke three ribs and lost two teeth. The local doctor pronounced these hurts incurable. We are told that when Priessnitz had applied his wet bandages and had drunk water in abundance for twelve months, he was fully restored—except presumably for the teeth. Neighbours, and then paying patients, flocked to him, and in spite of professional opposition that led the doctors to analyse his sponges to find out what he was using, by 1843 he had over 1,500 patients and a bank balance of £50,000.

These figures may have become known; certainly Hydropathic Establishments multiplied. Water was applied to the patients in a growing variety of ways. At one early establishment we see a Rain Bath for Medical Purposes: the patient is standing in a shallow brick pit surrounded by a boarded screen. On the ceiling above is a nozzle, with a tap worked by a cord led over pulleys. The doctor stands above and solemnly operates the cord to earn his fee.

It is no rare thing to see a subject who at this first shower betrays actual terror, shouts, struggles, runs away, experiences frightening suffocation and palpitation; and it is not rare to hear him say, after a few moments, 'so that's all it is'.

The Rain Bath, despite this terror, sprayed the patient through a rose, and had a much milder impact than the Douche Bath, in which a large-bore and powerful icy jet was directed at the ailing parts. Where the main water pressure was inadequate, a high overhead tank could be used: Dr. Wilson's tank at Malvern gave a drop of 20 feet, and a strong hat had to be worn to protect the head. There

The Rain Bath for Medical Purposes

were two painful accidents. One nervous lady stood on an unauthorised chair to reduce the drop, but when the deluge was released it broke the chair. A gentleman who took the Douche in winter was stabbed in the back by an unobserved icicle that had formed on the spout, though fortunately he was already too numb to know that anything was wrong until he saw the blood.

An alternate hot and cold douche ('a very potent type of bath') was for some reason known as an *Écossaise*. There were countless other varieties, but undoubtedly the most tremendous douche of all time was that taken by a daring American lady, under Niagara Falls —not, it is true, under the heaviest cataract, but in a minor fall near enough to the main sound and fury to share its dramatic effect.

159

As well as in its rather suspect communal form in the Hummums, the Vapour Bath came back as the Private Steam Cubicle or the Recumbent Steam Bath. The idea was not new. A British patent of 1678 had given directions for building a one-man 'Sweat Bath'. Another inventor in 1756 had published details of his 'Sudatory'. This was a vertical wooden box made of planks hinged together, with a seat inside adjustable to the height of the patient, and a hole in the lid for his neck. A copper kettle holding 6 quarts fed steam through a 3-in. diameter pipe into the lower part of the back of the 'sweating chair'—the patient might be sustained in his ordeal by 'drink or cordials'. This bath was better, claimed the inventor, than the 'common suffocating stoves at the Hummums'. It must have had some merits, for it was still being offered in its essential form into the 1920's, alongside the advertisements for curing the drink habit or removing superfluous hair. One of the earliest of the newly-revived Steam Baths comes from the U.S.A., where it was patented in 1814, and it is surprisingly more complex than most later models. As well as an impressive boiler to feed steam in under the bed-cover, it has a kind of four-poster canopy with curtains forming a roomy 'steam

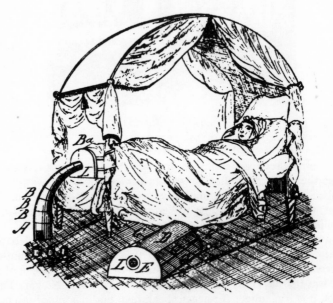

Steam Bed Bath, U.S.A., 1814

tent'. A German steam bath of 1832 is much simpler, being no more than a crude wooden box with a loose fabric cover raised on battens, in which the patient reclines. No steam generator is shown, and an ordinary domestic kettle with a steam pipe may have served. In an 1855 model, a bag contains the steam and the recumbent patient, and may be tightened round his neck by means of a cord. Sleeves in the bag permit his unaided escape, though he can not reach the boiler if it should get out of hand.

The *Encyclopaedia Britannica* of 1854 advises that

the vapour bath is infinitely superior to the warm bath for all the purposes for which a warm bath can be given. An effective vapour bath may easily be had in any house at a little cost and trouble.

(One wonders how many were 'all the purposes for which a warm bath can be given'.) Steam for these baths could be generated by many complicated devices, but *Britannica* goes back to the simple method of the mediaeval stews: a brick is heated in the oven and put

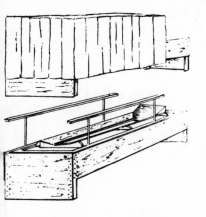

Steam Bath, Germany, 1832

Steam Bath, 1855

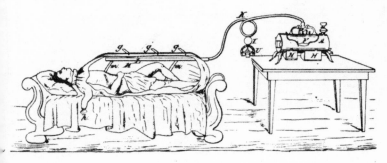

in a metal basin; water is poured over it to produce steam, and the patient, wrapped in a towel, sits on a chair above it.

For the Hot Air Bath, the simplest arrangement is the same as this, but with only a spirit lamp beneath:

Instantly the patient is seated the attendant places over him an inflammable robe, resting upon an iron framework which holds it a sufficient distance from the body to allow the hot air to come into contact with every part . . . a wet cloth is placed upon the head, which has a soothing effect and prevents the blood rushing to the brain . . . in 15 or 20 minutes the perspiration flows down the body in streams.

(The choice of an 'inflammable' robe will worry only those who do not know that in England then, as in America now, this meant the opposite of 'flammable'.) In Ewart's Hot Air Bath the patient sits in a horizontal wooden box, with his head emerging, his feet raised on footrests, with the lamp between them. He could hardly escape from this pillory in a hurry, and to overfill or overset the spirit lamp would be to invite a dreadful scene.

In the U.S.A., with a growing population going west, away from waterworks and drains, the portable steam or vapour bath, that could be ordered by mail, continued long in favour. Such was the

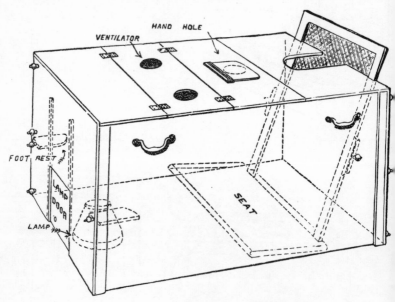

Ewart's Hot Air Bath, c. 1860

'Quaker' model, a fabric cylinder that cost 5 dollars, presumably excluding the chair. The good work was supplemented by 'Steam Doctors', who in the words of a more orthodox practitioner,

go about the country vexing, first the people's ears with strange jargon, and next their stomachs with cayenne and lobelia, and their skins with hot vapour.

This practitioner was the Philadelphia doctor John Bell, who set out to clean up the American way of life in 1850 with his *Treatise on Baths*. The full title of this work is rather longer:

DIETETICAL AND MEDICAL HYDROLOGY
A
TREATISE ON BATHS
including
COLD, SEA, WARM, HOT, VAPOUR, GAS
& MUD BATHS:
also,
ON THE WATERY REGIMEN,
HYDROPATHY
and
PULMONARY INHALATION;
with
A DESCRIPTION OF BATHING IN ANCIENT
& MODERN TIMES

163

Doctor Bell (in a style still echoed in the writings of professors in the minor American universities) sought what he termed 'a more satisfactory and harmonious doctrine of balneatory hygiene and therapeutics', continuing:

The author is not insensible to the ambition of giving greater vogue to the practice of bathing, under a belief that, if it were once to become general, it would contribute powerfully towards an increase of the public health, and of individual comfort and pleasure. It would be a step in the advance from physical to moral amelioration and progress.

. . . Bathing, after some fashion or another, may be considered as the gratification of an instinct common to all living nature; for, it is no fanciful idea to regard the aspersion of the vegetable kingdom, including all its tribes and varieties, from the cedar and the oak down to the humble and parasitical moss, by rain, atmospheric vapours and dew, as a modification of the bath, by which dust is removed, insects are destroyed, and fluid is furnished for the nutrition of the plant.

In other words, we should bathe, both for health and for pleasure; even the plants do it. The would-be bather is first invited to learn the various names for the bath, according to the part of the body to which the fluid is applied. Thus, it is a *semicupium* when only the lower half is immersed, a *pediluvium* when a foot bath, a *manuluvium* when a hand bath; a mud or earth bath is *illutation* or *illutamentum* or *lutamentum*. At this stage, a number of readers way out mid-West probably drop the whole idea, but those who press on learn that the duration of the *pediluvium* is, by some, declared to be ten minutes; by others, to last until the water acquires the same temperature as the feet, which will be from half an hour to an hour. The Cold Head Bath is practised by the patient reclining on a table, at one end of which is a 'vase' of water of a suitable size and depth to allow of his immersing first one side, then the other, and finally the back part of the head, giving about five minutes to each of the three regions.

'A Mode of Preparing a Warm Bath' starts with a procedure that may well have given rise to some suspicions.

An oblong case, of a size and form just sufficient to contain the human body, is constructed of deal. This is carried into the chamber of the patient, and there filled about one third with water of the requisite heat. A stout sheet is next laid over the aperture, and kept tight under the feet of assistants standing on each side. Upon this cloth the patient is placed, and, by slacking it, gently sunk into the water . . . The warm bath is alike removed from enfeebling depression and perturbating excitement, and it places the animal economy in a state of quietude most favourable to a correct balance of all its functions. . . . The hot bath induces acceleration of the pulse, some fulness of the head, and slight confusion of thought.

It will be noted that in all this, no actual bath is mentioned. Hygieia's American disciples seem to have managed with a table, a deal case, a bed-sheet and a vase or two. In 1851, a year after Bell's first edition, a real bath was indeed installed in the White House, but in the teeth of intense opposition. A survey in the 1880's revealed that five out of six dwellers in American cities still had no bath. In 1895 no New York tenement had one.

In England the virtues of the cold bath were increasingly preached, and the young especially were subjected to regular icy-cold plunges at all times of the year. A Dr. Marshall Hall had to protest against 'that mode of cold bathing which consists in the immersion of the infant over head in cold water'. At Winchester until about 1864 the scholars still had to wash at the old conduit in Chamber Court at all seasons—in the dark, in rain, snow and frost. Six brass cocks were fitted on the wall to serve them all; these often froze and had to be thawed out while the shivering queue waited. The monks were better off 500 years earlier. Until the end of the century a healthy young man who took his bath hot was thought effeminate.

Running water, when it came, at first seldom ran above the basement. It was soon piped to the kitchen sink, but had to be carried upstairs. The whole family might descend in turn to the stone-floored 'back kitchen' on Saturday night and bathe there, while a servant toiled to bring hot water from the 'copper'. The bath and wash-basin remained portable affairs, pieces of loose furniture or utensils without a room of their own. Wash-basins with taps, and baths with overhead shower tanks, are found in catalogues of the 1850's, but their water was raised by hand pumps.

Accustomed as we are to a more or less standard form of bath, we wonder at the astonishing variety of opinion in the nineteenth century as to what shape a bath should take. As they appeared in growing numbers, thanks to the improved manufacturing techniques and expanding markets arising from the Great Exhibition of 1851, their varieties included the Sponge Bath, the Full Bath or Lounge Bath, the Sitz Bath, the Hip Bath, the Fountain Bath, the Slipper Bath, the Shower Bath, and even the Travelling Bath that the clean-limbed Englishman took with him to the unplumbed Continent. All these were of sheet metal, individually made by craftsmen in copper, zinc or sheet-iron, painted or 'japanned'. Custom usually required the exterior to be plain brown, and the interior imitation marble. The finish did not stand up very long to wear and hot water, and the repainting of the bath was, until the fairly recent introduction of porcelain enamels, a regular and seldom successful operation, immortalised in the popular song

Out in the open, with everyone passing,
With everyone passing, and shouting 'What ho!'
She took her bath in the garden . . .
(*pause for laughter*)
. . . to paint it,
So what does it matter, I'd just like to know?

The Sponge Bath is circular, shallow, with tapered sides, a roll-edge, carrying handles, and a spout for emptying. It may have a little island in the middle which will be explained. More than adequate instructions were published for its use:

In taking such a bath it is desirable that the sponge be of large size, and it should be placed in the bath, charged with water, ready for immediate use. To obtain the fullest benefit in the most agreeable manner, the charged sponge, as the bather steps into the bath, should be lifted and carried quickly to the back of the head, which should be slightly inclined forward, so that the bulk of the water will run down the spine and back; the next spongeful should be almost instantaneously applied, leaning forward, to the top of the head, and the third, standing quite upright, to the chest; the arms and legs may then be separately treated: and if desire be felt for more, the application may be repeated to the back of the head and chest.

The use is recommended, in conjunction with the Sponge Bath, of a broad stool (heavily weighted at the bottom, to prevent the risk of upsetting) covered loosely with carpet, and high enough to reach above the level of the water when placed in the middle of the bath: the piece of carpet may be dried each day after use; or a Sponge Bath may be readily constructed with a fixed raised centre of metal forming a portion of the bath, the bather standing as it were on an island; the feet may thus at first be kept dry, and the preliminary shock received on the head and shoulders; persons who in despair had almost given up the Bath are by this means enabled to enjoy it without discomfort. Should the reaction after a Sponge Bath be very slow, it may be hastened by the previous addition to the water of a small wine-glassful of eau-de-cologne, spirits of wine, or spirits of any description, whiskey being perhaps best.

So that the learner might refer constantly to these vital instructions, there should surely have been an edition printed on waterproof paper, as was done later for d'Annunzio, who liked to read his own poems in the bath.

167

The Full Bath or Ordinary Lounge Bath is like a modern bath in form, about 5 ft. long, tapered, and may have a high back like an early motor car. There may be carrying handles and a drain cock. This type is rare until late in the century, probably because it takes so much water. The user is still described as 'the patient'.

Pour into the bath sufficient water to cover the body of the patient, with the exception of his head, when he lies upon his back. After he has lain in the bath for five minutes, his body should be well rubbed by the hands of a healthy bath attendant, and also by his own. It will add materially to the effect of the bath if the bowels or abdomen be gently, but thoroughly, manipulated and kneaded.

The Sitting or Sitz Bath is a square or oval vessel tapering to about 16 in. wide at the bottom, about 14 in. deep. It may have a perforated false bottom through which cold water will rise when poured down a funnel behind.

For this bath it is not necessary to undress, the coat only being taken off, and the shirt gathered under the waistcoat, which is buttoned upon it; and when seated in the water, which rises to the waist, a blanket is drawn round over the shoulders.

168

BATHS.

No. 1. The New Rising Douche Bath. Invaluable in cases of Prolapsus, Seminal Weakness, Piles, Generative Ailments, &c. Highly r commended. 30s.

No. 2. Shallow Baths, with plug, either in end or bottom ; beautifully enamelled. 25s. to 60s., according to size and strength.

Fig. 2.—Sitz Bath.

No. 3. Sitz Baths on Mr. Well's Special Design. No. 1, suitable for average sized people, 14s. and 16s. No. 2, larger size, for stouter people, neatly painted, 16s., 18s., and 21s.

No. 4. Portable Hot Air and Vapour Baths. From 22s. 6d. to 50s., according to completeness.

No. 5. Sponge Baths; strong, and well enamelled. 12s. 6d. to 35s.

No. 6. Foot Baths. 2s. 6d., 3s. 6d., and 4s. 6d., according to size.

STOMACH TUBES—For Washing Out and Cleansing the Stomach. These useful articles are invaluable to those who are troubled with Indigestion, Biliousness, Flatulency, Catarrh, Bronchitis, Asthma, Headache, and other Affections arising from Imperfect Assimilation of the Food, Colds, &c.

Composition Tubes, 3/-, 4/-, and 4/6.

Jacques' Patent Flexible Rubber Tubes, 4/-, 4/6, 5/-, and 5/6.

Silk Web Stomach Tubes, specially recommended, 6 -, 7/-, and 8/6.

Wired Rubber Tubes, 3/-, 4/-, and 5/-.

Let us hope that those who followed these instructions were not misled by the author's delicate omission of the removal of the trousers.

Dipping Sitz is a term applied to dipping the posterior part of the body a dozen or more times into cold water. This should be done slowly, and followed with friction. It is highly beneficial in cases of nervous debility or a relaxed condition of the generative parts.

The Sitz Bath is also recommended as 'very serviceable' for congestion of the brain. Admiral Beaumont describes Sitz bathing as 'hatching health'.

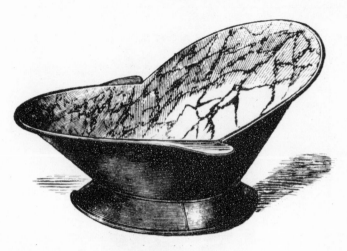

The Hip Bath, to become by far the most popular form, is oval, occasionally round, tapered downwards and with a base tapering outwards, a high back, a roll-edge and perhaps little elbow rests also serving as soap-dishes. Many older persons will still remember a hip bath set out in a cosy bedroom, on a waterproof bath-sheet, with brass or copper hot-water cans gleaming in the light of a good fire, and a thick towel warming on the fire-guard.

It is very beneficial in various forms of cholera, colic, liver complaints, diarrhoea, and disordered conditions.

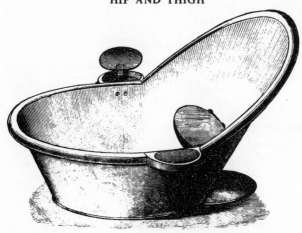

One model has a little seat half way up the back, to keep the rump out of the water; had it only been provided with foot-rests too, a bath could have been taken without getting wet at all.

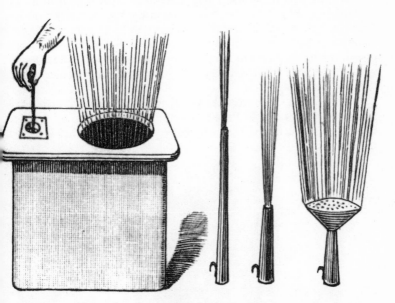

The Fountain Bath or Ascending Douche has variable sprays like a garden hose, giving an upward jet of water over which the patient sits.

171

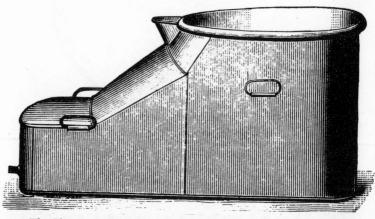

The Slipper Bath or Boot Bath (in France a *sabot*) is indeed like a boot, and being formed from about twenty pieces of sheet metal it is a complex piece of cobbling. Only the occupant's head and shoulders are seen; modesty and warmth are both preserved. In front of his nose is a little funnel for filling, and in the heel or toe is a drain-cock. In such a bath was Marat assassinated by Charlotte Corday. A variation that might well have been called a Wellington Boot Bath does not seem to offer any very comfortable position; it presumably has a raised seat; perhaps it was a special order for a customer whose legs had been amputated.

Confusion may arise from the fact that single baths of the ordinary modern kind, on hire at Public Baths, are there still called 'Slipper Baths'. Such, no doubt, they originally were; the term has survived the change.

DEATH OF MARAT
Although Marat spent hours in his bath, and thence conducted his correspondence, this was not from luxurious laziness; his 'sabot' bath offered the only relief from a painful skin disease

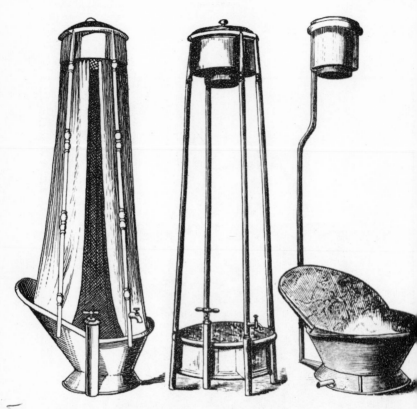

The Shower Bath is usually an Ordinary Lounge Bath or perhaps a Hip Bath with the addition of a small water-tank carried on three or four metal legs, one of which is a pipe up which the water, from a bucket, is forced by a hand pump. One model has only a small round bath of the same capacity as the tank. The user is warned that

nearly freezing water from a Shower Bath produces a feeling somewhat akin to what might be imagined to result from a shower of red-hot lead; the shock is tremendous, and the shower, if continued for any length of time, would assuredly cause asphyxia.

A shower like one of these may have been that one used by Disraeli, of whom his wife remarked that he had infinite moral courage but no physical courage, so that when he took a cold shower, she always had to come and pull the chain for him.

174

The Travelling Bath, with a lid and straps and perhaps with col-
lapsible wooden legs, is a cross between a big drum and a Sponge
Bath; there was even a Travelling Hip Bath. At first of metal, the
Travelling Bath might be had later in 'gutta-percha', and could no
doubt be folded up and put on the rack of the railway carriage.

Papier-maché was popular for some of the smaller types of bath;
it was light and cheap, and serviceable as long as the waterproof
finish lasted. But a combination of unobserved cracks and a heavy
bather must sometimes have led to a spectacular disintegration.

175

It may be wondered why there is, in all these instructions for use of the bath, no word of soap. But the bath taken for mere cleanliness, and not to cure anything, was still a special variety, and is separately described, as the *Soap Bath:*

The application of the Soap Bath is very simple: the bather is armed with a large lump of good ordinary yellow soap, and a loose washing glove (the white and very soft 'Turkish' is pleasantest, or if a hard glove be preferred, the 'Baden' may be used) big enough to come over the wrist, and standing *découvert* in front of the washing basin containing *hot* water, quickly and vigorously covers the body from head to foot with a thick and abundant lather . . . the soap application may take about three or four minutes at most; the very robust may use tepid water, but most persons will find warm or hot more agreeable; and more delicate persons whose finger-tips have the unpleasant habit of turning numb and white upon the application of cold water, will find in the Soap Bath an agreeable means of bathing perhaps otherwise unattainable . . . Drying and dressing may be comfortably gone through in front of a fire, the use of which for this purpose is a positive advantage, and will by no means, as might perhaps be imagined, tend to enervate or enfeeble. Scented soap must be avoided, especially for using habitually and liberally over the whole body. . . .
The Soap Bath may be commenced at any period of the year, and if children are induced to take it as a treat, rather than from any other point of view, they will soon become as partial to its use as their elders.

In these words, written in 1880, we see the beginnings of a new school of thought. Although the soap must not be scented, the bath may be taken warm or even hot; even in front of a fire; even by a healthy person. It may be enjoyed rather than being endured. The icy ordeal is not obligatory. Within two generations it will be rare; the twilight of the British Empire will fall upon steamy air. Nevertheless the editor of *The Boys' Own Paper*, well into the new century, repeatedly preached the virtues of the cold bath in his 'Answers to Correspondents'. In this feature only the answers are given, and the questions are tantalisingly left to be deduced from them, but it does seem that some of these correspondents must have confessed, like Aldred, to 'worldly thoughts'. The editor considerately adds to one of these cold bath prescriptions, 'but do not make more splash than you can help, so as to give the servant trouble'.

XIII

Was Health Worth It?

Bath Becoming Shabby — No Mixed Bathing — Dame Wakefield Indignant — Dr. Granville Not — He Looks Like a Lobster — Perfect Bliss in a Moorish Bath — David Urquhart — The Jermyn St. Bath — The Russian Bath — The Persian Bath — Hydros — The Bath Tariff — Intoxicants and Smoking — Mutton Chops in the Bath — Professor Wells of Observatory Villa, Scarborough — The Wet Sheet Pack — Dangers Of Amateur Use — Town Councillor Saved From Death — The Electric Bath — Mischief Done With the Wet Girdle — Patients' Struggles — Bulwer Lytton on Man's Duty — Dr. Wilson's Wet Sheet Deters a Suicide — Heals the Blind — Man of 80 Walks Uphill — Mr. Bradley Looks Mottled — Water Babies versus Wine and Beer Babies — Tarts in the Bedroom — The Atmospheric Cure — Dr. Rikli's Strange Costume

I N the early nineteenth century Bath had passed its heyday and was becoming a little shabby, but the waters had not lost their powers: about 1811 a lady was brought there *in extremis*, 'with all the frightful symptoms of death upon her', including 'despondency, sighing, swooning, singultus and convulsions, with an universal atrophy', and within six weeks she was dancing in the Town Hall. Hygieia had indeed taken over from Venus, and 'Wanton Dalliances' were out. Mixed bathing was barred early in the nineteenth century. Dame Wakefield, who boasted of twenty-two years' service as an attendant in the ladies' department, was indignant:

What harm did it ever do, or could it do, to see the nice dear creatures go down the steps out of their private undressing rooms, and enter the bath with their bathing-wrappers, made of rich stuff and fashionably cut, down to the feet and hands, and fastened to the waist,—their hair gathered up under a very elegant *coiffe*,—walk up through the water to shake hands and exchange morning salutations with the gentlemen of their acquaintance already in the bath, attired in the very pink of fashion? One might as well object to their walking together, or meeting and greeting each other in the Grove, in dresses not very far different. There they are immersed in air—and here they are immersed in water. Of the two, the latter is the more decent element, as it is not quite so transparent.

Or, in the words of Anstey's *Bath Poetical Guide*,

> Oh! 'twas a glorious sight to behold the fair sex
> All wading with gentlemen up to their necks,
> And view them so prettily tumble and sprawl
> In a big smoking kettle as big as our hall.

But Dr. Granville in 1841 approved of the new rule as 'infinitely less onerous to the feelings of delicacy and propriety which have ever characterised our fair countrywomen'.

Dr. Granville scientifically tested the effects of the Royal Hot-Baths, having first spent twelve hours on his legs, unfed since breakfast, perambulating the city.

On plunging the thermometer into the centre of the bath I found it marked 113½°, and the steam which rose from the surface of the water obscured the lamps with which the room was lighted. The attendant tried

178

to dissuade me from entering the bath at that degree of heat; he had never witnessed such an experiment, and felt sure it would do me great injury. To calm his apprehension, I promised to proceed cautiously, and requested him to be in immediate waiting in the adjoining apartment, that if he heard me call he might rush into the bath-room, and let the cold water flow into the bath.

I immersed the feet and legs into the water by descending three steps, but the sensation of positive scalding made me quickly retreat. In a few seconds I repeated the trial, and brought the water to my knees, by getting to the third step again: it was then bearable. I descended further so as to bring the water as high as the lower margin of the thorax, and found it again scalding and painful for a few seconds, during which time the whole part of the body exposed to the water had become of a fiery red, as I could distinctly perceive through the beautifully-transparent semi-greenish fluid. . . .

Encouraged, I proceeded lower down, and brought the water as high as the breast. My breath became suddenly short, and I felt seized as if I had plunged suddenly into very cold water. It was but for an instant; and I again found the temperature comfortable. I now lowered the thermometer deep into the middle of the stream, and saw it marked 114° of heat. My pulse beat heavy, full, round, and 100 times in the minute. There was a slight increase of noise in the ears; the head felt intolerably hot within, but neither heavy nor throbbing.

I let myself down fairly into the body of the water, floating between the surface and the bottom of the bath, the head alone being unimmersed. By this time the colour of a red lobster pervaded the whole surface of the body to an extraordinary degree; but, strange to say, instead of the skin of either the limbs or the fingers feeling, as it does when one plunges the hands or feet into ordinary water of high temperature, crisp and corrugated, the reverse was the case, for it felt quite smooth and soapy.

After 10 minutes' plunging, I tried the pulse once more, which had by this time risen to 115. My temples throbbed violently, and the singing in the ears had become louder; yet I was not sensible of any disturbance in the region of the chest. The bath-man was now summoned to let in cold water. . . .

and after thirteen minutes more, we are not surprised to hear that the Doctor cannot describe the satisfaction he experienced on reaching the dressing-room. He went on sweating profusely for half an hour, and his skin when dry at last felt 'as smooth as satin-paper'. The charges, he adds, are moderate: 30s. for thirteen baths, attendants included, and only 6d. more for any invalid or perfect cripple who should require the contrivance of a crane and pulley to let him in and out of the bath. In the Pump Room 'the volcanic stream is distributed to the subscribing invalid' at half a guinea a month, 3s. 6d. a week or 4d. a glass.

The Queen's Bath of 1597 had been demolished, revealing the circular Roman bath underneath. The King's Private Baths, above the Old Public Baths, date from 1788, and the Cross Bath, a 'cheap public bath' by the City Architect, Mr. Baldwin, from 1796, the date also of the Grand Pump Room. The New Royal Private Baths of 1870 included anew the Chair Bath, with an arm chair attached to a crane by means of which the helpless could be ducked. In 1888 the baths comprised three springs, the Grand Pump Room and five establishments for public and private baths.

The Illustrated London News of 24 April 1858 shows a picture of a 'Moorish bath' and gives a traveller's description of it.

The day was hot, the narrow streets were burning in the glare of noon. The prospect of the hot bath was not very inviting. I opened the door of the first apartment. All around it was a raised platform covered with mats on which lay several bathers in the state of profound repose. I was mounted on a pair of wooden clogs. The bath attendant rubbed and pinched and pulled every limb and joint of my body. He knelt upon my stomach so that I could hardly breathe, wrenched my arms and legs. . . . Having been pinched and poked and pressed sufficiently, this genius of the bath lathered me from head to foot and took in hand a huge glove, with which he proceeded to thrub me with the most lively animation. The amount of matter he managed to peel off the crust of the body is certainly surprising. Having been well soused in cool water, I was softly wiped and dried by another attendant. This done, he wrapped me up from head to feet in soft towels and led me to the outer apartment, the air of which seemed very much like that of an ice-house. I sank exhausted on a divan. And now it was ecstatic enjoyment, it was elysium, nothing seemed wanting to perfect bliss.

The terrors and joys of the Turkish Bath in London were similar, when with medical support they again revived. The most famous of all, the Jermyn Street Bath, was built under the direction of David Urquhart, a diplomat who had lived among the Turks and advocated their method. Urquhart incidentally found that in 1856 the Isle of Rathlin, off Ireland, still had mediaeval stews, used especially at the local Fair.

The Russian Vapour Bath, usually in a log hut with an open hearth, a pile of red-hot stones and a tub of water to produce steam, followed by a dousing with cold water, a plunge in a stream or a roll in the snow, still popular today in Russia and Scandinavia, never found favour in England, even when snow was plentiful. The mis-called 'Russian Bath' of a later date is only a private steam cabinet.

An English traveller of 1863 describes the Persian bath as a mixture of the Turkish and the Russian. He does not seem altogether to like the Persians or their bath. He takes the bath because, he says,

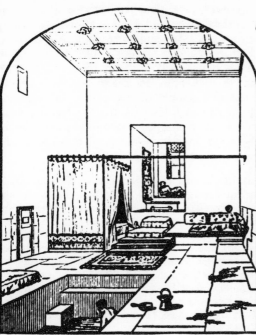

Urquhart's Design for
a 'Turkish' Bath

The Russian Bath

Europeans after having lived some time in Persia become dilapidated, like the Persians themselves, and there is always something dilapidated about the Persians. Men and horses, houses and walls are never quite sound: there is always a crooked tumble-down look about them. A trumpet is blown early in the morning to announce that the bath is hot, and many use this fact as a pretext for sitting up all night carousing, lest they should oversleep and not hear the summons. When a physician asks his patient if he has not been indulging too much in the warm bath, he means to enquire whether his nerves have not been shattered by debauchery. Folks go to the bath not to get washed, but to hear the news of the town.

If they seriously wanted to become clean—an idea, by the way, which never appears to enter into the imagination of anybody in Persia—the bath would be the very worst place they could go for the purpose. For, there exists an extraordinary notion among bathmen that a certain quantity of water can never become dirty. The bath, therefore, which is merely a huge tank filled with steam, and a reservoir for water, becomes glutted with abominations, and the water becomes as thick as pea-soup. Rats and black-beetles and horrible insects crawl about there; yet, inexpressibly filthy, foul and abominable as their baths are, the Persians watch over them with jealous care; the populace would probably rise in insurrection if a Christian were allowed to bathe there. They believe that it even cleanses from the impurity of sin.

He manages nevertheless to take a bath; his skin is peeled with a currycomb so thoroughly that for days he can scarcely bear the contact of his shirt.

To be sure the bald part of my head looks like a lump of gingerbread; but what hair I have appears not only to have been painted, but varnished too. Having resigned myself passively to the bathman, I find that he has also played wonderful tricks with my eyebrows, and with my nose, and with my ears. My beard looks like that of a youngster of twenty-three. I am astonished at my juvenile appearance, when I survey it in a greasy looking-glass.

. . . When I sum up my sensations calmly, I find I have been a considerable gainer, for although I am well aware that I came out of the bath much dirtier than I went into it—as one does also from a Turkish bath taken in Turkey—and that it has cost me two hours, part of my skin, and a headache, yet I know all the news of the town, and the bathman has flattered me so adroitly that I go away with a satisfactory idea of my own importance.

From about 1860 Hydropathic Establishments multiplied and flourished at the spas and seaside resorts. Many of them are still known as 'hydros', but they use less water than they did when the Bath Tariff offered more items than the Bill of Fare:

Arm Bath: Bran Bath: Dripping Sheet Bath: Dry Blanket Pack: Dry Blanket Pack with Wet Flannel Pad: Dry Rubbing or Friction Bath: Douche Bath: Electric Bath: Foot Bath: Foot Pack: Gargling Bath: Head Wash: Hot Blanket Pack: Hot Blanket Pack with Head Bath: Half Sheet Pack: Hot Wet Flannel Pad: Hot Air Bath: Lamp Bath: Leg Bath: Leg Pack: Mud Bath: Nose Bath: Ordinary Wet Pad: Plunge Bath: Rain Bath (weather permitting): Sitz Bath (Cold, Tepid, Hot, Running or Dipping): Hot Sitz and Back Sponge: Sand Bath: Slime Bath: Stomach Pan: Sulphur Bath: Towel Bath: Vapour Bath: Wet Socks (sic): Wet Sheet Pack: Wet Compress: Wet Girdle: Wet Bandage: Wet Head Cap: Wet Dress Bath.

All the makings indeed of a healthy holiday, but with attendant evils noted by one critic:

It is to be regretted, however, that at some of these establishments, beer and other intoxicants are provided and smoking permitted, and we also sometimes hear of the head physician prescribing bacon. . . .

The practice of eating a mutton chop, beef steak or drinking hot and strong tea or coffee while in the bath, or as soon as taken, is a very bad practice.

The critic is 'Professor' R. D. B. Wells, Phrenologist, of Obser-vatory Villa, Scarborough, who in *Water, and How to Apply It, in Health and Disease* (*c*. 1885) warns of the risks run by the amateur dabbling with this dangerous liquid.

While lecturing in a populous district where smallpox was raging, I applied the water treatment successfully to several hundreds of cases. After this, some of the inhabitants thought themselves clever enough to apply the wet sheet pack to themselves. One poorly and venturesome lady actually damped a sheet in cold water, wrapped it round her, and then went to bed! This, of course, gave her a cold, and made her so much worse that she was glad to call in our aid. We then gave her a wet sheet pack in the proper way, and she was surprised to experience such a wonderful change for the better in a very short time.

On another occasion a Town Councillor with small-pox, voted incurable by three doctors, was put out of danger by Wet Sheet Packs and Cold Wet Cloths round the neck. The family doctor, intervening, had the cloths wrung out in *warm* water, causing an instant relapse, and had not the Professor's assistant returned and removed the warm cloths just in time, 'it is questionable if the patient would have survived'.

It would be unwise to give a vapour or shallow bath to a man who had a very moist skin, inasmuch as more moisture would be added—and the system is already overcharged with it—and thereby throw it still further out of balance. Again, when the patient has a very moist skin, he will throw off electricity too rapidly, and thereby debilitate the body, and render it negative to surrounding influences.

In the Electric Bath the patient sits in a shallow warm bath with one pole of 'the magnet' in contact with the feet (perhaps wrapped round a toe) while the other pole is applied to various parts of the body through the medium of a wet sponge. The battery, we are warned, should be of superior make, as otherwise the shocks will be too severe and interrupted.

When the head is overheated with brain work, a Foot Bath at bed time is beneficial, but the toes must be kept in motion. When a Shower Bath is taken, a waterproof covering should be worn on the head. Mischief has sometimes been done by continuing the Wet Girdle too long; in some cases it has been worn day and night for months. But the Ordinary Wet Pad may, thanks to the Professor's specially-designed apron, be worn all day long. The Hot Blanket Pack will be more efficacious if the patient fasts for from one to three days, not taking a particle of solid food, but drinking freely of warm water. 'Where there is strength enough to withstand it', the Pack may be followed by a Pouring Sheet Bath. It is a common error to bind the Wet Sheet Pack too tightly:

The Wet Sheet Pack

Sometimes patients have felt so oppressed in this way, that in the absence of the attendant they have in their struggles to free themselves rolled off the bed on to the floor.

'Professor' Wells seems to have strayed into the late nineteenth century from the Domenicetti period. The Marx brothers' script-writers do not seem to have come across his book.

Sir Edward Bulwer Lytton, who took the Water Cure at Malvern, could not speak too highly of the Wet Sheet:

I think it is the duty of every man on whom the lives of others depend to make himself acquainted with at least this part of the Water Cure. The Wet Sheet is the true life preserver.

Dr. Wilson of Malvern found that the Wet Sheet, provided one managed the struggle to apply it, would deter a man bent upon suicide. One of his patients who was nearly blind was by its use restored to sight. A venerable old Quaker, who came to Malvern unable to walk across the room without palpitation, was met in his eightieth year 'firmly walking uphill' by Mr. Lane, who took the Cure in 1846 and wrote a book about it.

On our return we met Mr. Bardley airing himself after his blanketing, and looking very mottled. He said he felt relieved and very happy.

Mr. Lane assures his readers that 'a Hydropathic skin is your true flannel waistcoat, and the best protection against the elements', but we notice from the illustrations that he and his teenage son, who do a lot of walking and climbing, invariably wear top hats and tail coats too. A couple who have been childless for eighteen years take the Cure at Malvern and produce two fine babies within twelve months; Dr. Wilson admits that many who have never been Water Patients have babies, but points to the marked distinction between 'Water Babies' and 'Wine-and-beer Babies'. Dr. Wilson's only real worry was that insubordinate patients used to smuggle forbidden items, such as tarts, up to their rooms. Pastry was bad enough, but Miss Asplin, who tried to take up a warming-pan, had to be sent home in disgrace.

With the Water Cure went the Atmospheric Cure, involving the unheard-of practices of sun-bathing and exercise in light clothing. The special costume devised by Dr. Rikli in the 1860's, with open-necked shirt and shorts, looks like that of any hiker today, or with the addition of a little feathered trilby hat, like any cartoon German or Austrian, but it was wildly revolutionary then.

Bath nowadays is a little sad and very strange. To tour the main group of buildings is to relive one of those inconsequential surrealist dreams: elegant picture galleries lead suddenly to a sort of roofed bomb-site where Roman remains mingle with iron pipes, white glazed brick, exhibition displays, and runnels of rusty water. A prison cell exudes suffocating steam. Only goldfish use the main bath, but a few pale youths furtively carrying towels are shepherded by worried mothers through odd-shaped corridors with Edwardian tiling and coloured glass, where waitresses carry trays. It is not quite a museum, not quite a ruin, not quite a tea-shop, not quite a hospital, not quite an underground station under construction, but something of all these at once. In the Pump Room a trio plays the lightest of light music to an audience of crocheting gentlewomen and white-haired squires in plus-fours reading *Country Life*, many in a state resembling death, though one will sometimes stir and tiptoe among the ferns to draw a glass of the fecundating waters. The scene is watched from the doorways by little knots of almost disbelieving tourists.

XIV

'The Benison of Hot Water'

*Methods of Heating Bathwater — Hot Stones — Pots and Pans
— The Copper — Range and Boiler — Self Heating Baths — The
Bath Comes To Rest — Water Heating by Gas — Defries' Gas
Heater — Bos Piger's Disappointment — Fearful Explosion in
Kitchen — Piped Hot Water — The Gas Bath — Maughan's
Geyser — Sugg's Boiling Stream Therma — Jennings' Solid Fuel
Geyser — Strode's Geyser — Evolution of the Foolproof Geyser
— The Multipoint Pressure Geyser — Newest Refinements*

Wᴇ have seen hot bath-water produced, first by heated stones, then in pots or pans heated on a fire or stove, then in the basement or outhouse in a 'copper', an open hemispherical metal container built into a brick stove, that still survives in its primary use as a clothes-boiler where the laundry and 'Launderette' are remote. The next stage is a boiler built into the kitchen cooking range. A 'Combined Fireplace, Boiler and Cast-iron Oven' comes out as early as 1806:

By drawing a damper either of them will heat by the fire that is used in common. The boiler should have a tube with a brass cock which projecting into the kitchen gives hot water whenever wanted.

In 1814 the *Morning Post* carries an advertisement for a Patent Steam Kitchen and Range that, besides cooking, will supply constantly from 1 to 14 gallons of boiling water, 'all performed by the use of one small fire'. The next stage is not to pipe hot water to the bath to avoid carrying it, but to carry cold water to the bath and heat it on the spot. Hot bath-water, therefore, appears before cold water is piped upstairs. An article in *The Magazine of Science and School of Arts* in 1842 records that

many copper and tin baths have lately been constructed in London, with a little furnace attached to one end, and surrounded with a case or jacket, into which the water flows and circulates backwards and forwards till the whole mass in the bath gets heated to the due degree,

adding the very necessary caution that 'when the proper temperature is attained, the fire must of course be extinguished'.

Such an installation was shown at the Great Exhibition in 1851; the English heating engineer had reached the same stage and was finding the same troubles as his French counterpart of a century earlier. The bath is filled and emptied by hand, and solid fuel is used, so that the disposal of coal smoke or charcoal fumes offers a problem.

188

The domestic 'bathroom' of 1846 is merely a recess in the bedroom. It has a cold water supply, but no hot water, no waste and no overflow. No mat is shown to protect the carpet, although the bath is nearly ready. There are two small towels, but the large fabric is a sheet ready for use as a Wet Sheet Pack. We have no illustration of the state of affairs just after such a bath

Warm Bath Apparatus, manufactured by Tylor and Son, London.

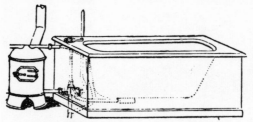

A warm bath by means of circulation may be obtained in about half an hour. The apparatus is portable, and can be used in any room where there is a flue.

(At Deane, Dray, and Deane's, King William Street.)

Portable Bath Heater, 1850

Henry Cole's *Journal of Design* of 1850 recommends a heated bath 'that can be used in any room having a stove pipe', implying that the bath must now settle down permanently in one place, but portable heated baths continue to be made without this problem being faced. An early portable bath by Jennings has the local heating arrangement as well as an overhead shower filled by a hand pump.

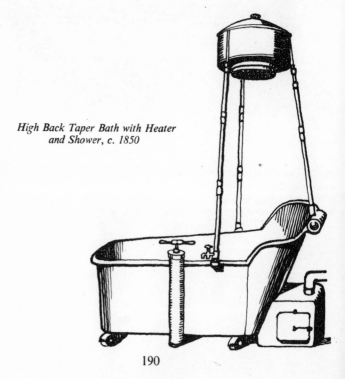

High Back Taper Bath with Heater and Shower, c. 1850

190

As the cold water supply pipe 'like a wounded snake, drags its slow length along' and reaches the upper floors, and the cold water tank finds its accepted position in the frost-liable roof space, the nomadic bath comes to rest in a room of its own, anchored by a complex network of piping. This room is at first a converted bedroom, needlessly large, with wallpaper, curtains and furniture quite unsuited to a damp steamy atmosphere. Where no room, or even a recess in the bedroom, is to spare, the folding bath, looking something like a wardrobe when not in use, offers a solution.

Although the gas supply was commercialised in 1812 with the granting of a Royal Charter to the Gas Light & Coke Company, it was not until the 1850's that gas was used for heating water. Attempts were made from time to time to devise gas-heated vessels for bath water, but seemingly by persons who did not understand the basic principles. Defries, in *The Journal of Gas Lighting* for 1850, claimed that his 'Magic' Heater would give a hot bath in six minutes with 2*d*. worth of gas. It had no boiler but was set under the bath.

Every establishment ought to be provided with this indispensable requisite for the comfort and preservation of life. Its limited cost placing it within the means of all, and its simplicity within the management of a child.

A bunch of flames playing on a metal bath, capable of heating even ten gallons of water in six minutes, without a flue, can hardly have fulfilled these claims.

In 1867 a correspondent of *The Gas Journal* signing himself *Bos Piger* described how he called in

two conjurors, friends of mine, to advise upon and fix a gas-stove, close to the bath cistern, in aid of the kitchen fire.

The boiler was 1 ft. in diameter and 18 in. high, and circulated water to a hot storage tank. Below it was a gas ring, and the whole was enclosed in a copper jacket. Poor *Bos Piger*, whose kitchen fire already must have given inadequate hot water, was again disappointed, for

immediately it is lighted the whole house, top and bottom, stinks of half-consumed gas, and there is a considerable accumulation of caked soot at the bottom of the jacket, fallen off from the bottom of the boiler.

By this time the hot water boiler built into the kitchen range was fairly common. Many such installations seem to have been put in about 1869, for in December of that year there was a whole series of fearful explosions in kitchens. In one such near Manchester, where

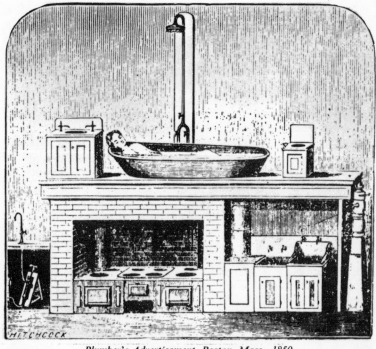

Plumber's Advertisement, Boston, Mass., 1850

a lady and her daughter were seated at the fire, the former was killed
and the latter scalded; the ceiling was brought down, the walls battered, the windows smashed; a cat and a dog crouching by the fire
also died. The cause of these bangs was the practice of putting the
hot water storage cylinder near the top of the house, and with no
safety-valve. The cold water tank aloft freezes unobserved; hot water
is drawn off and not replaced; the boiler runs dry; the cold tank
thaws, and cold water enters the red-hot boiler. If the boiler is kept
very small, and the cylinder is put downstairs, a freeze-up aloft will
merely stop the flow at the taps; the cylinder and boiler remain full,
and there can be no explosion. The plumbers learned these simple
truths, at the expense of their customers and their dumb pets.

The hot water cylinder, near the range, was vertical, of copper,
and uninsulated. No really orderly sequence of development in hot
water plumbing can be traced and dated. There might still be no cold
water tank aloft to provide pressure, but a hand pump, perhaps in
the basement so that a servant could work it, could send both cold
and hot water up to the bath and basins. A plumber's advertisement

192

of 1850 from Boston, Mass., shows such an arrangement, but its interpretation offers some puzzles. The lady in the bath appears at first glance to be on the kitchen mantelpiece, but when the conventions of the diagram are grasped it will be seen that she is on the upper floor. The bath is rather long, or she is rather short. Bath and shower have mixing taps. The pump, seen below on the left, seems to be portable, and its flexible tubes suggest that its connections may have been variable. It can hardly have kept up an adequate pressure for very long in the system. The 'closet' seems to have no water-flush or waste, but to be a mere 'commode'. There are presumably bath and lavatory wastes under the floor, unseen. The object on the extreme right invites interpretation.

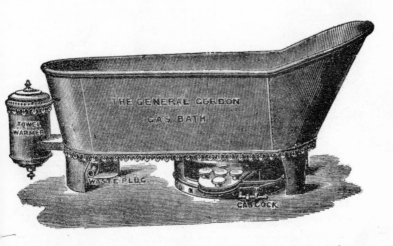

In the Gas Bath proper, combining bath and gas heater, the gas burner swings out for lighting, and as in Defries' 'Magic' Heater it plays directly on the underside of the bath. It is turned off, one hopes, before entering; if not, the error would not be made twice. Ewart's were very early in this field, and the Gas Bath was later to come to full flower (in 1882) in their 'General Gordon', with its charmingly-detailed Towel Warmer. A cheaper device was the portable heater, connected to a gas lighting bracket and placed in the bath water itself; this type survived until the 1920's, or revived, as the 'Otazel'.

In 1868 Benjamin Waddy Maughan produced the first gas 'Geyser'. (Every reader of Cleasby and Vigfusson's *Icelandic Dictionary* knows that the word *geyser* is the same as *geiser* or *geisir* and is from the Icelandic *geysir*, gusher or rager, from the verb *geysa*, a derivative of *gjosa*, to gush, meaning an intermittent hot spring.)

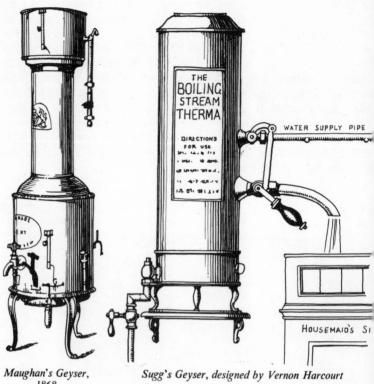

Maughan's Geyser,
1868

Sugg's Geyser, designed by Vernon Harcourt

The Science Museum has a fine Maughan geyser, its casing painted as green marble and its fittings in shining brass. It stands on three dainty legs, and with its Royal Arms has something of the charm of an early Great Western locomotive.

Sugg's 'Boiling Stream Therma' followed, and at the same time Ewart's were making geysers for both gas and oil. Jennings went one better with a model that used coal, coke or wood, but at some sacrifice of speed:

The Geyser is filled either direct through a tap or by hand, as most convenient, and the fire having been lighted, the apparatus must be left 15 to 30 minutes, according to the temperature of water desired

—but this is hardly a true geyser, which is properly a water-tube boiler, in which a water pipe of copper or wrought iron takes a winding, spiral or zigzag course through a cylinder containing burners.

194

In some early geysers, however, the water seems merely to have trickled or sprayed downwards among the flames, without a pipe, for as late as 1896 an advertisement for the 'Tubal' claims, as if for an unusual feature, that 'the water does not come into contact with the gas, and is thereby kept quite fit for culinary and drinking purposes'. The water often entered a funnel fed by a separate tap of the ordinary kind. The performance of some of the early geysers compares quite well with that of today: at the Crystal Palace Exhibition of 1882, a unit made by Strode, using less than 100 cu. ft. of gas an hour, could raise 80 gallons of water an hour through 60° F. The first geysers had no safety devices, and it was important to follow some complicated instructions.

The evolution of the foolproof geyser may be seen as a long conflict between the engineer—who thinks it stupid and unnecessary to make a bang, because he has operated the thing a hundred times in the workshop without trouble—and us, the ordinary silly public (especially those of our womenfolk who get flustered by things mechanical), who seem to defeat every new safety-device in turn, making a bang in some remaining way that has not been guarded against. At first the odds are heavily against us: there is no pilot jet, and if our match goes out, we probably omit to turn off the gas before striking another; there is a hearty bang accompanied by clouds of green powder. If the water supply fails, or if we turn it off before turning off the gas, the geyser emits steam and molten solder and

Geysers, 1890–1896

195

finally disintegrates. The engineer, muttering about stupidity, makes the burner carriage swing out for lighting, and interlocks the gas- and water-taps to enforce a safe sequence of operations. We are less nervous now, but we can still re-light the gas in full flow after the carriage is closed, with as good a bang as ever. The angry engineer introduces the pilot jet, but if this goes out, matters are much as before. Exasperated, he devises a thermostat control that will not admit gas to the main burner until the heat of the pilot jet itself has opened the gas tap. He also makes the gas flow dependent on the water flow, and defies us to make a bang in future. We hardly ever do, but ways still remain to us: if we mishandle the taps and momen- tarily turn off the pilot jet, then on again, the thermostat may not cool in time to deny us one explosive cylinder-full of gas, that will bang well if we mistime the match accurately; if the pilot jet goes out, even its modest flow may in time provide us with the requisite charge. But these are rare combinations of events, and it is probably only a few obsolete geysers that still occasionally give those unmistakeable distant crumps that can be heard in the quiet of a London Sunday morning. A new generation is growing up without our inborn fear of the 'gusher or rager'.

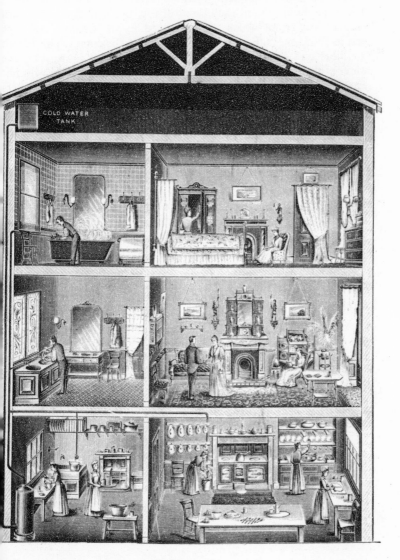

COLD WATER TANK

'Sectional drawing of House, showing the distribution of Hot Water in the various rooms, by means of High Pressure Circulating Gas Boiler'

The first geysers, not being made to stand internal water-pressure, had to be duplicated at each point of use, but with the multi-point pressure geyser, of which the long-lived Ewart 'Califont' was the first in 1899, one sufficed to feed every tap in the circuit. This layout sacrifices one great advantage of the geyser, that by heating the water at the point of use it avoids the waste of filling a pipe with hot water. It can lead to angry competition between would-be users of different taps; if the kitchen-maid runs hot water into her sink, the master of the house can find his bath cold. This trouble was overcome by the gas-heated storage tank, which in its later forms embodies a thermostat. The newest refinements are temperature controls for the geyser that can be set for summer or winter use, or that can supply water at any temperature from tepid to boiling.

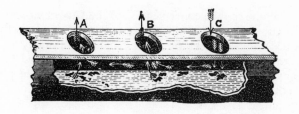

XV

'The Subject is a Peculiar One'

Thin Iron Ovens — George Jennings' Crusade — Did Crystal Palace Need Lavatories? — Halting Stations — The Annus Mirabilis of the Water Closet — Hopper Closets — One Lav For The Rich And One For The Poor — Twyford's Washout Closet — Hellyer's Optimus Closet — Mann's Syphonic Closet — Jennings' Pedestal Vase — Two Odd Tests — Twyford's Washout Pedestal Closet — Bostel's Washdown Closet — Famous Collections Of Ancestral Seats — The Rev. Henry Moule's Earth Closet — Earth In The Bedroom — The Wonderful System At Woodroyd — Fever Attacks The Wrong Class — The Prince Of Wales Has Typhoid — Eloquence Of Stevens Hellyer — A Retort To Punch — Gruesome Examples — Dr. T. Pridgin Teale Raises A Stink — Mr. Horsfall's Sporting Butler — The Sanitary Engineer's Nose — Ingenious Devices — The Plumber Always Hangs The Bell — His Status And Wages — The Conservancy System — Water Closets Even In Manchester

I N 1858 the 'public conveniences' of the City of London were few and foul; at best 'thin iron ovens', such eyesores that the nearby householders or office tenants had many of them removed, and by their continued protests prevented the erection of anything better. The subject was indelicate, and the problem was not admitted to exist. Then came George Jennings. Jennings had installed public lavatories in the Crystal Palace for the Great Exhibition of 1851, and 827,280 (or 14 per cent) of the visitors had paid for their use. The Official Report states that

No apology is needed for publishing these facts, which . . . strongly impressed all concerned . . . with the sufferings which must be endured by all, but more especially by females, on account of the want of them.

In spite of this, when the building was re-erected at Sydenham, it was strongly urged on grounds of economy that lavatories should be excluded. Jennings was told that 'persons would not come to Sydenham to wash their hands'. Fortunately for the public he won the argument, and his installations produced a revenue of £1000 a year. In 1858 Jennings was crusading for the provision of 'conveniences suited to this advanced stage of civilisation' in place of 'those Plague Spots that are offensive to the eye, and a reproach to the Metropolis'.

I know the subject is a peculiar one, and very difficult to handle, but no false delicacy ought to prevent immediate attention being given to matters affecting the health and comfort of the thousands who daily throng the thoroughfares of your City . . . the Civilisation of a People can be measured by their Domestic and Sanitary appliances and although my proposition may be startling I am convinced the day will come when Halting Stations replete with every convenience will be constructed in all localities where numbers assemble. Fancy one of these complete, having a respectable attendant, who on pain of dismissal should be obliged to give each seat a rub over with a damp leather after use, the same attendant to hand a clean towel, comb and brush to those who may require to use them. A shoe black might do a shining trade, as many go with Dirty Boots rather than stand exposed to the public gaze.

The startling part of the proposition was that the 'Halting Stations' should be put underground. The wall space was to be covered with railway time-tables and lists of cab fares, less alarming than the notices of today. Jennings offered to supply and fix the appliances free of charge, together with 'respectable attendants, capable of understanding and answering a question', if he were allowed to charge a small fee: it is to Jennings that we owe the expression 'spending a penny'.

My offer (I blush to record it) was declined by Gentlemen (influenced by English delicacy of feeling) who preferred that the Daughters and Wives of Englishmen should encounter at every corner, sights so disgusting to every sense, and the general Public suffers pain and often permanent injury rather than permit the construction of that shelter and privacy now common in every City in the World. . . . I could give you particulars as to the opposition I have had to encounter from individuals, who if measured by their obstructive policy, one would suppose that they themselves never required any convenience of any kind.

Not until the seventies was the battle won. It is mainly to George Jennings that we owe the present wealth of those 'conveniences which nature demands, and every decent man, and every thoughtful mind approves'.

'Nothing can be more satisfactory than a good water-closet apparatus, properly connected with a well-ventilated sewer,' says Mr. Eassie (*Sanitary Arrangements for Dwellings*, 1874), and although we sense some overstatement in this unqualified chapter-opening, we see what he means.

1870 was the *annus mirabilis* of the water-closet. Until that year, the old 'Bramah' had served well enough, its only rival being the Hopper Closet. In the Long Hopper Closet, a conical pan was flushed by a thin spiral of water, but in Hellyer's words 'with such a twirling motion, that by the time it has twirled itself down to the trap it has no energy left to carry anything with it'. The area to be

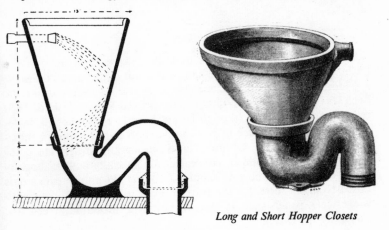

Long and Short Hopper Closets

cleaned was far too big, and even the Short Hopper Closet was little better. Made in two pieces of fireclay, it was all too easy and cheap to manufacture; one was advertised as 'suitable for Prisons, Mills, &c.', though another was to be had in two qualities, 'The Cottage' and 'The Castle': one lav for the rich and one for the poor. Hellyer suggested that instead of destroying the thousands already made, they might be used by market gardeners for protecting rhubarb from frost.

About 1870 Mr. T. W. Twyford of Hanley noticed that he was getting 2s. for the pottery part of a 'Bramah', whereas the brass- and iron-founder was getting from 20s. to 50s. Twyford conceived the all-earthenware closet of 'washout' form. In the Washout Closet a very shallow bowl holds an inch or so of water (unless this has evaporated), and although the flush may empty the bowl, it loses most of its force in doing so and 'gravitates through the trap in a most unselfish kind of way', says Hellyer, 'taking little or nothing with it'. Despite this, Twyford's sales soon reached 10,000 a year, and washout closets were lately to be found still in use.

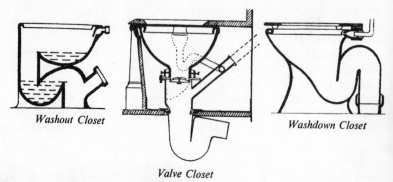

Washout Closet   *Washdown Closet*

Valve Closet

In the old 'Bramah' valve closet there were three defects: the flushing operation failed if the handle was not pulled up all the way; a seldom-used closet could lose its water-seal by evaporation; the flush was noisy. Hellyer's 'Optimus' Improved Valve Closet of 1870 overcame these defects (though he did have to recommend a taste-fully-lettered little notice in gold on ivory to ensure proper handle-pulling) and many an 'Optimus' is still in working order. The complex ironmongery could be concealed in a specially-built mahogany casing or a small mahogany cabinet, standing like a throne on a dais raised to make room for the trap, or even in an elegant chair with wickerwork panels such as Dent & Hellyer's 'Moreton' at £24 10s. excluding the closet.

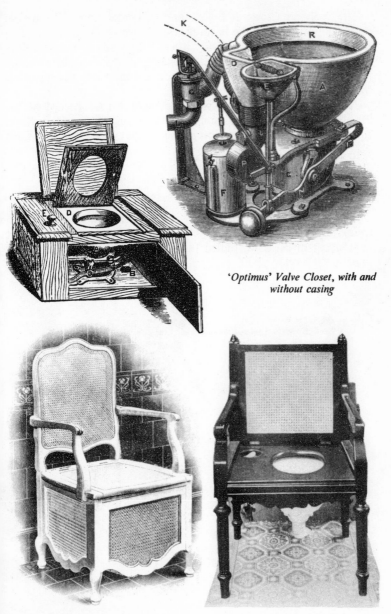

*'Optimus' Valve Closet, with and
without casing*

*The 'Moreton' Chair Enclosure
for the 'Optimus'*

*'Closet of the Century',
1900*

The Syphonic Closet of J. R. Mann also dates from 1870. Before its handle was pulled, the basin held some water. On pulling, a fast flush was followed by a slower after-flush, while syphonic action kept things moving. Here at last was a closet that did not advertise its every use loudly throughout the house, and in this respect better than many closets of today.

Jennings' 'Pedestal Vase' won the Gold Medal Award at the Health Exhibition of 1884, being judged 'as perfect a sanitary closet as can be made'. In a test, it completely cleared with a 2-gallon flush

10 apples averaging $1\frac{1}{4}$ ins. diameter
1 flat sponge about $4\frac{1}{2}$ ins. diameter
Plumber's 'smudge' coated over the pan
4 pieces of paper adhering closely to the soiled surface.

A simpler test was that improvised by Mr. Shanks who, when trying a new model, seized the cap from the head of an attendant apprentice, thrust it in, pulled the chain, saw it go and cried out happily, 'It works!'

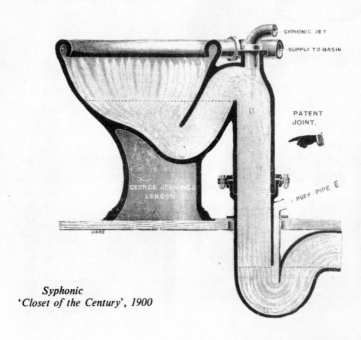

Syphonic
'Closet of the Century', 1900

Epic' Syphonic Closet, 1897

In 1885 Twyford came out with the washout 'Unitas', claimed as the pioneer of pedestal closets and certainly close in date to Jennings' 'Pedestal Vase'. These began the sound practice of fixing closets open and exposed, without woodwork, whereby joints can be examined and corners and dirt-traps abolished. That ominously-named device the 'save-all tray' became less necessary.

The invention of the washdown closet, the current form, about 1889, has been claimed for Mr. D. T. Bostel of the well-known Brighton firm, who incidentally have a good collection of ancestral seats. This is simple and efficient and can be made in one piece. It has the minimum of exposed surface, and the force of the flush passes through it unimpeded. Like earlier types, it may need a 3-gallon flush rather than the 2-gallon limit imposed by most Water Boards, and it can be shockingly noisy. But essentially the problem has been solved, 293 years after 'Ajax'.

The valve type of flush needs no interval for filling between each use, but has come up against objections as to possible waste of water, or even demands by the water authorities that it be fed through a meter. A device which automatically closes the valve, after a pre-determined quantity of water has been discharged, has overcome these objections.

205

'Blue Magnolia Design', 1895

'Raised Acanthus Pattern', 1895

'Pedestal Lion' Closet

'The Lambeth', 1895

'Mulberry Peach Decoration', 1890

'*The Dolphin*', 1882

'*Lowdown Suites*', 1895

The British Transport Commission has some fine historic water-closets in its keeping. A collection kept by Messrs. Dent and Hellyer has unfortunately been lost by enemy action. A more tragic and less forgivable disaster has befallen the examples lovingly gathered by Dr. Parkes, the founder of the Parkes Museum, entrusted to the then Royal Sanitary Institute. When this body changed its title to The Royal Society of Health, these irreplaceable milestones of sanitary history were cast out, being sold for £5 to an amateur who has been unable to house them, and they are irretrievably dispersed. Let it be hoped that any such heirlooms may in future be bequeathed to the Science Museum, and room found there for an ordered display. There is too often a lag of about twenty years between the destruction of some piece of seemingly worthless junk, and the realisation that it had a serious historic value.

One side-issue in this field remains, that dates a little further back. The sermons of the Rev. Henry Moule are forgotten, for all their godliness, but his name lives in the adjacent annals of cleanliness. In 1860 he invented the Earth Closet. In its simplest form, a wooden seat has a bucket beneath, and above at the back is a hopper filled with fine dry earth, charcoal or ashes. When a handle is pulled, a layer of earth falls into the bucket, which is emptied at intervals. Chemists assure us that the end-product quickly becomes sterile and inoffensive. Special little stoves have been made for drying the earth beforehand, but the oven or the greenhouse will serve. If ashes are used, a sieve may be introduced above to salvage useful cinders. An ingenious variant is operated by a slight movement of the seat when the user rises; to strangers the unexpected sound-effect can be startling. 'In sick rooms, this method of distribution of earth may be found objectionable, as more or less vibration follows the rising, and this is apt to disturb the nerves of a patient'. It is considered unsporting to use the earth closet on occasions when a gentleman need not sit down.

Despite the chemists' assurances, we raise an eyebrow on reading in an advertisement of 1900, by a famous firm best left unnamed; that

Commodes, or portable closets, are serviceable in bedrooms and nur-series; whilst in sick rooms, in hospitals, and in infirmaries they are in-valuable. Each commode is fitted with an earth reservoir, with apparatus suitable for delivering the earth, and with a pail. The reservoir must be filled, and the pail must be emptied, as often as may be requisite.

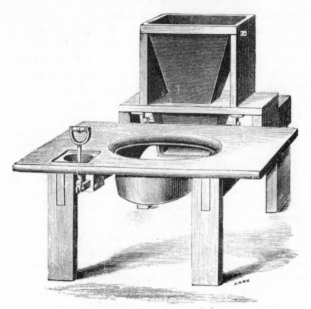

Moule's Earth Closet, 1860

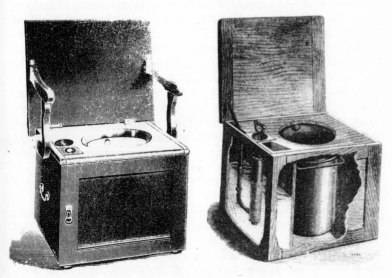

Portable Water Closets 'with pump and copper pail', 1882

On Moule's simple basis, a Mr. Garrett in 1898 erected and patented perhaps the most elaborate system ever contrived for a private house. At 'Woodroyd', near the Bigsweir railway station on the G.W.R. in Gloucestershire, the indoor earth closets were arranged one above another, with a large shaft behind them outside the wall, down which the output fell into a bin at ground level. Fresh earth was raised by means of a sack, pulley and rope up the same shaft. According to Mr. Garrett the bin needed to be cleared out only once a month. From the bin the used soil was put into a 'special barrow', and from this into the scavenger's cart. The cart too was 'special', being arranged to tip the contents into a 'special' railway truck for removal into the country. At this point we lose sight of any further trans-shipments. We have no picture of 'Woodroyd' with its strange appendage, nor any figures for the cost of shaft, bin, cart and railway truck; none for the wage bill, and no record of Mr. Garrett's first interview with the Great Western, at which, one suspects, the *privately-owned* railway truck was their idea. Never in the history of sanitation was so much done by so many for so few.

The new sewers, the public baths, and the general awakening to the effects of bad plumbing, did not bring quick results. The last cholera wave in 1866 was by no means the end of trouble. Enteric fever and typhoid went on, and worse still, began to show a puzzling disrespect for high-class plumbing and high-class persons. Death seemed unwilling to bless the squire and his relations, and keep us in our proper stations. An outbreak of enteric fever at Worthing almost exclusively attacked the well-to-do houses on the higher levels, where the water-closets were indoors, and spared the poorer quarters. Typhoid ignored the nearby village schools when it decimated the young gentlemen of Bramham College. A worse breach of propriety was to follow. In 1871 the Prince of Wales, afterwards King Edward VII, stayed at Londesborough Lodge near Scarborough. Soon after his return to Sandringham he went down with typhoid. As the Earl of Chesterfield, who had been of the party, and the Prince's groom, both died, it was obvious that this house was the source. Now national feeling was really aroused: the heir to the throne had nearly paid the penalty for the Countess of Londesborough's drains.

In a new crusade, Stevens Hellyer was to the fore. The Prince, recovering, took interest, and made one much-quoted remark, that may with all due respect be thought a shade unconvincing: if he could not be a prince, he said, his next preference would be to be a plumber. In the august premises of the Royal Society of Arts, Hellyer eloquently exhorted his fellow-craftsmen:

Lying there in those strong arms of yours, slumbering in the hardened muscles, resting in the well-trained fingers and educated hands, lies the health of this leviathan city!

If Hellyer could not have been a plumber, he might perhaps have been a poet. Nor would he have plumbers laughed at. Stung by the old joke about the plumber who goes back for his tools, he retaliated:

How the public are amused when *Punch* or the Press try to 'show up' the British Workman! But I fancy the B.W. gets the larger fun out of such caricaturing. The man who writes the skit from his easy chair only wants pen and ink to do *his* pretty piece of work; but the poor plumber . . . may want any *one* of a kit or chest of tools that would raise a sore on the back of a donkey to carry them, and would need the air to be scented with carrots to get him along.

Hellyer's are the best of the early textbooks. In vivid language, with some naïve but telling pictures by his own hand, he condemns the old brick drains ('elongated cesspools'), drains too large to be self-cleansing, drains running uphill or through right-angled joints, waste-pipes acting as flues for sewer-gas, unventilated soil-pipes, untrapped fittings, inefficient traps, pan closets, washout closets, drain vents near windows; all with current examples as gruesome as anything that had been uncovered by the reformers of a generation before. He objects to many of the new water-closets that are 'springing up like mushrooms'—a vivid picture this—and complains with some reason of having to wait eight minutes before a second flush can be given.

'A Dangerous Water Closet'

211

Dr. T. Pridgin Teale likewise raised many a stink. Like Hellyer, he drew his own pictures, for his *Pictorial Guide to Domestic Sanitary Defects* (1874), and they repay examination if only for their irrelevant details. In the original, the arrows representing smells are given an added horror by printing them in a poisonous-looking green. He reveals the interesting fact that surgical operations were often conducted in the patient's home: in one such, the surgeon refused to cut open the patient until the soil pipe under the floor was cut off from the drain. Another such pipe was so rotten that it 'crumbled like shortcake', and the kitchen walls 'ran with black damp'. And Mr. J. Horsfall found his butler angling for rats in an open cesspool in the basement. This sort of thing, says Teale,

is probably the cause of headache, sore throat and depressed health to many a cook, kitchen-maid and butler, and perhaps indirectly leads, in not a few instances, to the use of those treacherous self-prescribed medicines —spirits and beer.

The sanitary engineer of those days needed above all a delicately trained nose. He literally smelled out trouble. He learned to distinguish between ammoniacal, dank and putrid scents, 'neither are all dangerous smells of a foul nature, as there is a close, sweet smell which is even worse'. He was aided in his horrid hunt by such

Rats, and the tale they tell

T.P.T. Inv.

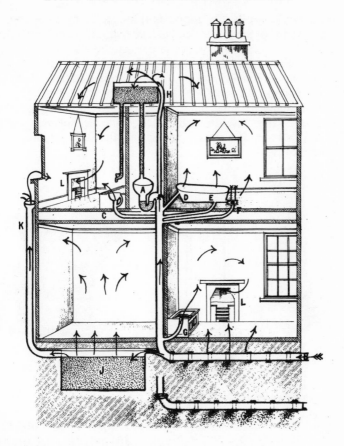

A. *Water-closet in the centre of the house.* B. *House drain under floor of a room.*
C. *Waste-pipe of lavatory—untrapped and passing into soil-pipe of W.C., thus
allowing a direct channel for sewer gas to be drawn by the fires LL into the house.*
D. *Over-flow pipe of bath untrapped and passing into soil-pipe.* E. *Waste-pipe of
bath untrapped and passing into soil-pipe.* F. *Save-all tray below taps untrapped and
passing into soil-pipe.* G. *Kitchen sink untrapped and passing into soil-pipe.* H. *Water-
closet cistern with over-flow into soil-pipe of W.C. thus ventilating the drain into
the roof, polluting the air of the house, and polluting the water in the cistern, which
also forms the water-supply of the house for drinking and washing.* J. *Rain-water
tank under floor, with over-flow into drain.* K. *Fall-pipe conducting foul air from
tank fouled by drain gas, and delivering it just below a window.* L. *Drain under
house with uncemented joints leaking; also a defective junction of vertical soil-pipe
with horizontal drain; the drain laid without proper fall.*

(*Dr. T. Pridgin Teale, 1878*)

ingenious devices as the peppermint test, whereby oil of peppermint was inserted in the system so that a distant fault could be sniffed out also by the smoke test which in one baffling case revealed that the elusive effluvium was coming down the chimney into the living room by 'Drain Grenades' and by 'Pain's Drain Rockets' which 'emit a dense volume of smoke with a pungent smell'.

It is significant that many patent gadgets were being offered to disinfect water-closets, usually on the principle (still accepted today) of 'a smell to hide a stink'—recognising the fault but curing only its symptoms.

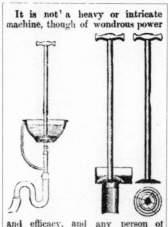

It is not a heavy or intricate machine, though of wondrous power

and efficacy, and any person of ordinary intelligence can, by the use of it, remove the most obstinate obstruction in a few moments.

It is not an expensive article, as the price, 7s. 6d., brings it within the reach of all, and it will last for years

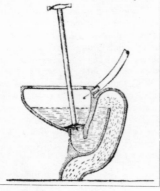

Burnet's Patent Smoke Drain Tester.

DIVISION D. **WADE'S "GOVERNMENT"** SECTION '2'
AUTOMATIC DISINFECTOR.
PATENT NO. 9276.

Will Fit any Flush Tank or Cistern.
It disinfects ALL the Flush Water passing to the Drains.
Lasts 1 to 3 years, during which it requires no attention.

REQUIRES NO FIXING
SIMPLY PLACED IN
AN ORDINARY CISTERN

CAN BE RE-CHARGED IN
A FEW MINUTES   TO DRAIN

FOR USE IN W.C.'S & URINALS,
ALSO IN
ROAD WATERING VANS,
Ensuring a Regular Supply of the Disinfectant.

A Hygienic protection. practically securing the user an immunity from Typhoid, Diphtheria, and other Diseases due to Drains.
UNFAILING. ECONOMICAL. CONVENIENT.

The plumber included in his duties the hanging of bells (a bell-wire could conveniently be run down inside a vent-pipe if the entry and exit holes were concealed), and later the installation of the new electric bells, with their row of wet cells containing sal-ammoniac, 'blest product of the camels of the East'.

His status and his wages were rising:

	Per day	Per week	Hours per week
1837	4s. 6d.	27s.	54, working all day Saturday.
1844	5s.	30s.	,, ,, ,, ,, ,,
1855	5s. 6d.	33s.	61½, till 4 p.m. on Saturday.
1861–73	6s. 4d.	35s. 10d.	58½, till 1 p.m. on Saturday.

The hourly rate in 1837 was 6d., in 1896 10½d., in 1918 1s. 5½d. It is now about 4s. 9d. In 1879 Dent and Hellyer were awarding every deserving employee an imposing Certificate of Merit:

215

The 'conservancy' or 'pail-and-tub' system of disposal continue
into the twentieth century. At Rochdale in 1896, with the exceptio
of 750 water-closets to the 'better houses', the authorities use
wooden pails made from petroleum casks (*sic*) cut in two. An empt
pail was left when the full one was collected. Specially coloured pai
were supplied to houses where infectious diseases had been notified
In Manchester the pail system did not replace the middens until 1871
but we read that by 1898 *'even in Manchester'* water-closets wer
being introduced, and the pail system fully retained only in Hul
Rochdale, Warrington and Darwen.

By this time the Angel of Death was behaving more rationally: a
Leicester the typhoid in Navigation Street was five times worse i
the houses *without* water-closets, which was as it should be.

XVI

Plumber's Progress

Further Evolution Of The Wash-stand — Running Water Arrives Upstairs — Cast Iron Comes To Full Flower — Face Showers — Tip-up Basins — One-piece Fireclay Basins — The Pedestal Basin — Further Evolution Of The Bath — Vanderbilt's Simple Bathroom — Houses Designed With Bathrooms — Panelled Enclosures — The Hooded Bath — Ten Controls — Stained Glass Windows, Curtains, Wallpaper And Carpets — The Cast Iron Bath — Paint Troubles — The Porcelain Bath — The Metal Hooded Bath — Showers — The Bathroom Becomes More Practical — It Shrinks — Influence of the American Hotel Plan — The English Inn — Porcelain Enamels — Fireclay Baths — Mass-produced Baths

WE last saw the wash-stand in its form of about 1780. About 1830 a new form emerged, much larger, rectangular, virtually a table, with a marble top and a wooden or marble upstand. The bowl was larger, not sunk, and in it stood a jug holding at least a gallon of cold water. There were china soap-dishes with loose perforated linings, a sponge-bowl similarly lined, tooth-glasses, water-bottles and even a container for false teeth. On a lower shelf was a china slop-pail with a shallow china funnel and a lid; perhaps a wickerwork handle; there might be a small china foot-bath. In a double bedroom all these items would be duplicated. Later the upstand was enlarged to become a tiled splashback, with a shelf above it. The hot water was brought in a brass or copper can wrapped in a towel. The housemaid who brought this upstairs had a wash-stand of her own, but of painted deal, and with oilcloth in place of marble.

This type survived until running water reached the bedroom, some time after about 1870, and indeed survives still where plumbing has yet to penetrate. When running water came, the bowl became a fixture, sunk into the marble top, and to hide the supply and waste pipes the whole was cased in with panelled and carved mahogany. Mechanisation by no means killed decoration. Combined with the tiled splashback, a framed mirror and towel rails, the wash-stand could be most imposing. It might appear in the bathroom as well as in the bedroom or dressing-room. Fixed though it was by its plumbing and by its very massiveness, it was still treated as a piece of furniture, even to the extent of keeping an open dust-trap, between vestigial 'legs' in what could have been a solid skirting recessed for toe-room. Like most furniture of the time, it asserted itself boldly, rather than being subordinated to its room as the unobtrusive furniture of the eighteenth century had been. It was expensive, and was meant to look so.

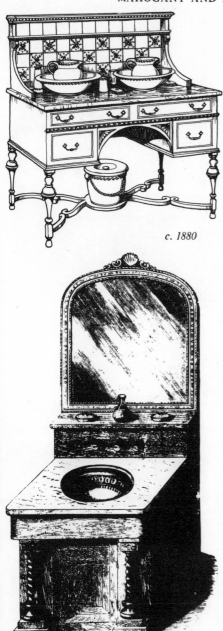

c. 1880

1877

1882

1882

Cabinet Lavatory, 1890

Lavatory, 1877

Bathroom by Hellyer, 1877

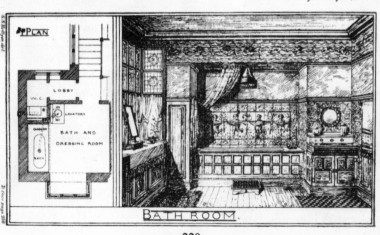

1895

Lavatory by Hellyer, 1877

Lavatory Combination, 1900

1895

221

The taste for such solid emblems of caste and wealth duly perco-lated down to the middle classes. It is a stock principle in marketing new luxury goods, to exhaust your thinner top layer of wealthy buyers, while rarity-value keeps the price up, before reducing the quality a little, and the price a lot, for the middle-class market. The monumental wash-basin of about 1880, in spite of the flood of machine-made mouldings, tiles and ornaments pouring from the factories, did not lend itself to quantity-production in its original materials, but the answer lay to hand in cast-iron. The astonishing possibilities of this material, well shown at the 1851 exhibition, enabled the wash-stand to be made twice as ornately at half the price. In the catalogues of Adams, Bolding, Froy, Jennings, Shanks, Twy-ford and others, cast-iron almost literally comes to full flower. The wood-cased wash-basin gives way to an open affair with the earthen-ware basin set in a cast-iron frame. Legs, brackets, towel-rails, shelves, mirror-frames and coat-hooks in ferrous filigree can be multiplied at the mere turn of a ladle. The styles, in so far as they have roots at all, range from Empire of a sort to *rococo* of a kind, but with more of Coalbrookdale than of Paris. The iron is unashamedly painted to imitate wood, marble, bronze or gold. Froy's fantasy in Late Marzipan style, complete with taps, comes out at £7. The cata-logues of from 1880 to 1900, rich with lithographic colour and gold, are a treasury over which, disapprove as we should, we pore with nostalgic pleasure.

A forgotten luxury offered in the nineties was the Face Shower that rose like a fountain from the bottom of the wash-basin. The tip-up basin, that emptied into a close-fitting lower container and needed no plug, was popular for some decades, but the container being out of sight became offensive.

About 1900, one-piece basins in whiteware or glazed fireclay, com-plete with flat tops and recesses for soap, previously deemed impos-sible to produce, replaced the composite forms, and were left exposed for easier cleaning. Such simple objects would formerly have been deemed fit only for servants' rooms. The overflow was sometimes made cleanable—a necessary provision not always made even today. Soon after 1900 the pedestal wash-basin, already available in cast-iron ('nicely bronzed or japanned') came out in fireclay: neat in appearance and easy to clean, though the latter advantage was often lost by over-ornamentation.

Lavatory basins in bedrooms were an uncommon luxury until about 1918, though as early as about 1870 Hellyer had graciously allowed that with proper plumbing and ventilation 'there seems little harm in granting the indulgence'.

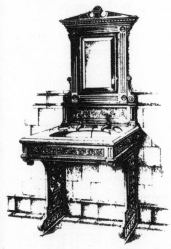

Cast Iron Lavatory Stands, c. 1900   *Pedestal Tip-up Lavatory, 1890*

George Vanderbilt's New York Bathroom, 1855

The bath, when it first comes to rest in a proper bathroom with running hot water, is a fairly simple affair of sheet metal, with no more than a little wood-graining or stencilled pattern by way of decoration. George Vanderbilt's New York bathroom of 1855, though it has that rare luxury a 'porcelain' bath, cannot be called ostentatious. All the piping is frankly displayed. The fittings are reduced to the essentials and compactly arranged. The forms are simple and functional. The bathroom is no more thought of as a fit place for display than the scullery.

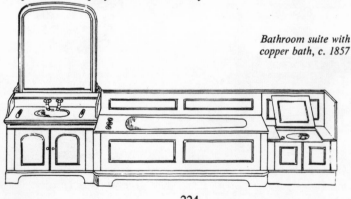

Bathroom suite with copper bath, c. 1857

224

1890

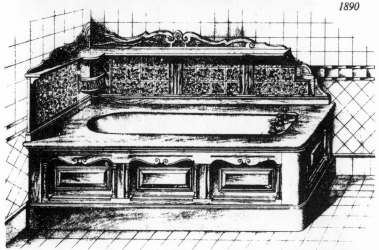

1895

By about 1880 things are changing. There is more money about. The bathroom offers a means of impressing one's friends with one's wealth and good taste. Like the gold-plated Daimler of a later period, it is more important as evidence of riches than as a useful piece of equipment. Although most bathrooms are still converted bedrooms, new houses are being designed with bathrooms, but these have the same extravagant floor area. The bath is encased in panelled wood-work, with an elaborate wood-framed tiled splash-back. More imposing still is the 'Hooded Bath', with its shower cabinet cased in carved mahogany worthy of a Renaissance altar-piece. Jealous of

225

Superior Bath Cabinet, 1899
Improved Spray Bath, 1882
Patent Oriental Spray Bath, c. 1890

the cabinet-maker's display, the plumber too runs riot. Ewart's 'Improved Spray Bath' of 1882 has no less than ten gleaming control handles whereby the happy bather can play at will on countless aquatic variations: hot or cold shower, spray, douche, wave, sitz or plunge. Froy's 'Patent Oriental Spray Bath', without connections, costs up to £147 7s. 6d., and this is merely a stock pattern; special designs are offered to architects' requirements and doubtless cost much more. Everything is very big and solid: the brass taps of Queen

Victoria's bath at Osborne look capable of operating the boilers of a
battleship. Bathroom windows are usually of stained glass and have
heavy tasselled curtains. Jennings' luxury bathroom of the 80's still
has patterned wallpaper, but tiles soon become more popular for
walls and floor. Shanks' luxury bathroom of 1900, though it has a
tiled floor, still has a large carpet. Its stained glass windows consort
oddly with its Ionic pilasters of Siena marble. Wash-basins, sitz baths,
foot baths and bidets, all at dignified distances apart, and brass
towel-rails and gasoliers, complete the grand display.

Changes in the material of the bath itself have had little visible effect on the bathroom. The cast-iron bath came in about 1880; the pioneers were probably either Cockburn or the Carron Company. The ironfounders may seem to have been slow to produce this simple-looking object, considering that the first iron boat had been cast in 1740 and the first iron bridge in 1770. But no one component of boat or bridge presented such technical problems as the casting of a smooth one-piece bath without undue thickness and weight. The cast-iron bath was not at first a luxury article. The rim was flat, like that of the sheet-metal baths, to take a wooden top; the cast roll-rim came many years later. The iron was galvanised or painted; the paint was a source of trouble. For years manufacturers claimed in turn that they had at last, and for the first time, perfected an enamel that would stand up to the wear and the hot water. Many householders still found repainting necessary sooner or later, and some of them ran into trouble, as did Mr. Pooter of *A Diary of a Nobody:*

April 26. Got some more red enamel paint (red, to my mind, being the best colour), and painted the coal-scuttle, and the backs of our *Shakspeare* the binding of which had almost worn out.

April 27. Painted the bath red, and was delighted with the result. Sorry to say Carrie was not, in fact we had a few words about it. She said I ought to have consulted her, and she had never heard of such a thing as a bath being painted red. I replied: 'It's merely a matter of taste.'

He bought more enamel paint—black this time—and spent the evening of the 28th painting the fender, picture-frames, an old pair of boots, and his friend Gowing's walking-stick.

April 29. Bath ready—could scarcely bear it so hot. I persevered, and got in; very hot, but very acceptable. I lay still for some time.

On moving my hand above the surface of the water, I experienced the greatest fright I ever received in the whole course of my life; for imagine my horror on discovering my hand, as I thought, full of blood. My first thought was that I had ruptured an artery, and was bleeding to death, and should be discovered, later on, looking like a second Marat, as I remember seeing him in Madame Tussaud's. My second thought was to ring the bell, but remembered that there was no bell to ring. My third was, that there was nothing but the enamel paint, which had dissolved with boiling water. I stepped out of the bath, perfectly red all over, resembling the Red Indians I have seen depicted in an East-End theatre. I determined not to say a word to Carrie, but to tell Farmerson to come on Monday and paint the bath white.

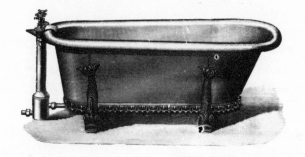

These illustrate a few of our Designs, but any scheme of Colouring can be adapted to the various patterns as may be desired. Special Designs on application.

Marble-veining was still the popular interior finish. The outside decorations became more and more elaborate, and by 1900 were being offered in almost as many varieties as wallpaper.

For the wealthier customer there was the porcelain crockery bath, made in one piece. It was durable and easily cleaned, but heavy and fragile for transport, and very cold to the touch until well warmed up by the water.

Just as the expensive composite wash-stand was followed by a popular model in cast-iron, so the mahogany hooded bath was followed, about 1900, by an all-metal counterpart. The shower enclosure becomes a semi-cylinder of sheet zinc, double-shelled, perforated inside to give a fine spray as an alternative to the overhead shower. A hinged door gives access to the pipes and fittings. The outside of the shower enclosure offers further scope for floral fun. Shanks' 'Independent' Plunge, Spray and Shower Bath, 'metallic enamelled' inside and japanned outside, 'fulfils the sanitary requirements of the age' and costs only £27 10s. complete.

No. 114A

This Bath fulfils the sanitary requirements of the age, in that there is an entire absence of wood enclosure, and the possibility of filth collecting behind that enclosure is completely obviated.

The Plunge Bath is of Enamelled Cast-Iron, with Shanks' Patent Waste and Overflow arrangement. The Spray may be of Enamelled Zinc or Copper, with an outer casing of Zinc enclosing all the Pipes, but with a hinged door for access to the Pipes and Fittings at any time. The Shower is of Brass, Nickel-Plated, and the Fittings are Shanks' Patent Eureka Fittings. The whole thing is perfect and complete in itself. Only three joints are to make, and the Bath is fixed. Great saving in Carpenter and Plumber work is secured.

This Bath is an improved pattern, with Circular Spray like the Eureka Bath, and with the Waste Pipe and Supply Pipes covered in at side of Bath.

The whole is enamelled and decorated in best style, and may be finished in various designs.'

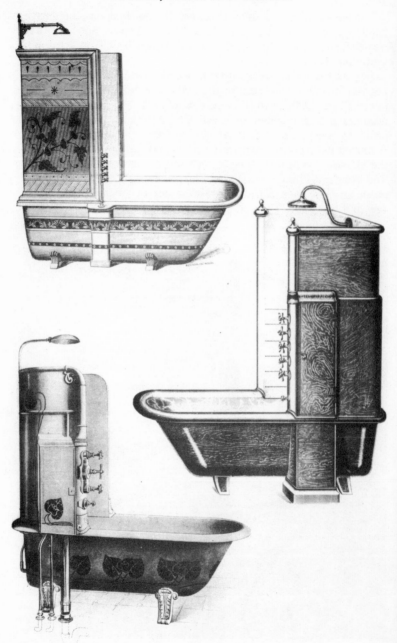

'*Model Bathroom*', 1902

At the same period, simpler showers are offered in two forms. Both forms have a ring carrying a cylindrical waterproof curtain. One form can be fixed over an ordinary bath; the other stands separately over its own shallow trough and waste. With such a shower, a large heated towel-rail and a wash-basin on tubular legs, polished nickel-plating becomes the keynote of some bathrooms. Woodwork, curtains and carpets disappear in favour of white enamel, tiles and marble. The bathroom is no longer a furnished room. It is still impressive but more practical.

233

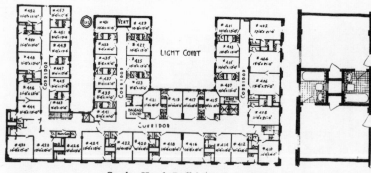

Statler Hotel, Buffalo, U.S.A., 1908

Space-saving in the bathroom, in its shrinkage from a great furnished room to a compact 'bath cell', had begun in the American hotel. In 1827 a Boston hotel had water-closets and bathrooms, though these were all in the basement. In 1853 the Mount Vernon Hotel at Cape May, New Jersey, had a bath with hot and cold running water for every bedroom. This costly idea was so slow to spread, that not until 1906 did the Ritz in Paris follow suit, and in 1908 the Statler Hotel at Buffalo made a big stir when it advertised 'A Room With a Bath For a Dollar and a Half'. Authorities were gradually persuaded that a properly ventilated and plumbed hotel bathroom need not necessarily be on an outside wall. The hotel plan began to include a block of two bathrooms and two cupboards, with a small light-well and pipe-duct, or cross-ventilated, between two bedrooms. Elevations became tidier as a result. American architects understandably criticise the English way of pattering in a dressing-gown along a corridor to the bathroom. Perhaps we do what we can; America has fewer mediaeval inns than England, and a lovable place like the Bull at Burford (its front modernised, in spite of angry protests, in the sixteenth century) might lose some diehard customers, and the special character that brings them there, if its good landlord somehow squeezed in a smart little bathroom to every twisted mediaeval bedchamber. To such diehards it is enough that the water is hot and the bathroom is clean, if not always vacant.

Soon after 1900, the bathroom became smaller, with all the fittings and plumbing on one wall. For a large house, architects began to prefer several small bathrooms instead of one large one that could be used only at agreed times by the children or the servants. The bathroom in even the largest house, excepting occasional costly fantasies, became essentially the same as that in the smallest new-built villa.

234

Ionic pilasters, Siena marble, stained glass and a carpet: 1899. Below, the 'Ribbon and Shell' design, 1899

The compact 'bath cell', 1908

Western civilisation had at last made up its mind as to what form of bath it really wanted: a solo tub about 5 ft. 6 in. long, parallel-sided or slightly tapered, with a roll-rim and with taps attached to the bath itself. The large and growing demand made a cheap standard product practicable. Sheet metal, needing skilled handwork, went out of use. About 1910 the cast-iron single-shell bath, enamelled on the inside only, and either painted outside or cased in, was in quantity production at a price the masses could pay.

At about the same time the problem of the finish was solved by the perfecting of 'porcelain enamels'. These are neither porcelain nor enamel. Strictly they are vitreous coatings akin to glass. Save for appearance, they could be as clear as glass, but white or tinted 'opacifiers' are added. The basic ingredients of the enamel are sand, lime and sodium carbonate. These are smelted together, quenched and milled to a powder. The bath casting is sandblasted, then kept red-hot while the powder is dusted over it, repeatedly and with re-heatings, so that the enamel fuses to the metal. A good porcelain enamel will expand and contract at the same rate as the bath and is almost indestructible.

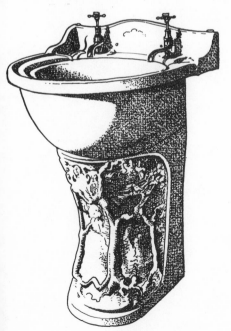

Art Nouveau in the Lavatory, 1907

The fireclay bath, the only serious rival to cast-iron, was formerly made by building up clay on a mould by hand. Correct thickness depended entirely on the skill of the operator. About 1906 such tricky hand-work was eliminated with the introduction of casting. The clay being liquefied by adding water and chemical salts, the fluid is poured into a porous plaster mould that absorbs the water. The result is a bath of perfectly uniform thickness and a maximum strength-to-weight ratio. The baths are then baked serially in tunnel kilns, like loaves. From about 1916 fireclay baths were being mass-produced. At about that date came the mass-produced double-shell cast-iron bath, porcelain-enamelled inside and out, the standard type today. Over and above the baths made for the home market thousands have been exported to the Orient, where a sad side-effect has come about: the wealthy having adopted the Western-style bathroom, the Hammams have lost all but their poorer patrons, letting their rich decorations and furnishings decay, and lapsing into squalor.

XVII

Toilet Sundries

Soap and Substitutes — Castile Soap From Castile — Soap
Taxes — Addison Approves an Advertisement — Dr. Johnson
Disapproves One — Suave, Sans Pareille, Vento's Italian Water
and Miss In Her Teens — Curious Cyprian Wash Balls — Trans-
parent Soap — Knight, Gibbs and Yardley at the Great Exhibition
— Pears Spend £100,000 a Year — Rowland's Macassar Oil —
Bear's Grease — Shooting the Bear — Razors, Shaving Soap,
Strops — The Hair Plane — Mr. Gillette's Motives — The Teeth
Brush — Cloth To Rubb My Mrs. Teeth — Tooth Blanch —
Tooth Soap — Turner's Dentrifice Fastens The Teeth — The
TAM HTAB — Towels — The Flesh Brush — Old Media for
Bathing — The Lunatic Fringe — Dr. Sanctorius' Bag Bath —
The Author's Improvements Thereon — Milly's Mechanical Bath
— Sheraton's Oddest Fancy — The Piano Bed — Folding Tubs —
Sanitary Conjuring Tricks — Brandy in the Lavatory — The Jeep
Bath

S OAP and its allied toilet products deserve a brief history. Soap
was unknown to the ancients. They did use various 'soap-
weeds' that are said to give an excellent lather. The Egyptians
may have added natron, a form of carbonate of soda, to their
washing water. Pliny says that the Gauls made 'soap' from goats' fat
and beech ash (potash), but that this was only for brightening the
hair (*rutilandis capillis*). The Roman mixture of oil and sand has
already been mentioned. Galen in the second century is the first to
mention soap for washing the person or clothes. In England it was at
first home-made, but it was manufactured commercially from the
fourteenth century. It was used mainly in the laundry; 'toilet waters'
in great variety were for a long time preferred for personal use. A
lady who used 'myrrh water' found it

good to make on lok younge longe; I only wete a fine cloth and wipe my
face over at night with it.

In the sixteenth century great quantities of Castile soap were
imported from Spain, while peace lasted. London soap boilers made
three kinds: speckled, which was the best, white, and grey which was
the cheapest. These were sold by the firkin, and buyers added their
own perfumes. The Soapers' or Soapmakers' Company was incorporated in 1638. By the time of Charles I the industry was sufficiently
important for him to be able to sell the monopoly rights; rival soapmakers had friends in Parliament, and soap may thus be listed among
the causes of the Civil War. The Roundheads did not favour undue
attention to the toilet, and Cromwell severely taxed soap. This tax
was lifted at the Restoration. By 1700 there were sixty-three soap
factories in London. In 1712 the tax was re-imposed at 1*d*. per lb. An
Act of that year laid down 'British standards' for soap.

Addison in *The Tatler* in 1710 quotes an advertisement submitted
to the paper, as a pattern of good copy-writing:

The highest compounded Spirit of Lavender, the most glorious, if the
expression may be used, enlivening scent and flavour that can possibly be,
which so raptures the spirits, delights the gusts, and gives such airs to the
countenance, as are not to be imagined but by those that have tried it.
The meanest sort of the thing is admired by most gentlemen and ladies;
but this far more, as by far it exceeds it, to the gaining among all a more

han common esteem. It is sold in neat flint bottles, fit for the pocket,
only at the Golden Key in Wharton's Court, near Holborn Bars, for
three shillings and sixpence, with directions.

Addison admires 'the several flowers in which this spirit of lavender
is wrapped up'.

Dr. Johnson, criticising advertisements in *The Idler* of 1759,
mentions with special scorn 'The True Royal Chymical Washball'
which besides ridding the skin of all 'Deformities, Tetters, Ring-
worm, Morphew, Sunburn, Scurf, Pimples and pits or redness of the
Small Pox' is warranted to 'give an exquisite edge to the razor, and
so comfort the brain and nerves as to prevent catching cold'.

Among the toilet items taxed under a Statute of George II in 1786
are powders, pastes, balls, balsams, ointments, oils, waters, washes,
tinctures, essences and liquors; among the toilet waters are listed
'Suave', 'Sans Pareille', 'Vento's Italian Water' and 'Miss In Her
Teens'.

Advertisement drawing by
Ernest Griset, 1898

In 1789 Andrew Pears began making his transparent soap, by dissolving ordinary soap in alcohol, distilling off the alcohol to produce a transparent jelly, and drying this slowly in moulds. From 1815 the soap tax was 3*d.* per lb., or about 100 per cent, but in 1833 the tax was halved, and in 1853 it was repealed by Gladstone, a measure well timed to the revival of the bath habit. England was then producing about 136 million lb. a year. At the Great Exhibition of 1851 there were 727 exhibitors in the soap and perfumery section, about half of them British. Knight, Gibbs and Yardley were there; the last-named, as Yardley and Statham, won an award for their Brown Windsor Soap, and the very same cake was shown again 100 years later at the Victoria and Albert Centenary Exhibition. Pears pioneered new advertising methods on a huge scale; their expenditure rose from £80 in 1865 to over £100,000 a year. Another widely advertised product was Rowland's Macassar Oil, which appeared about 1793, and had the distinction of being mentioned in *Alice Through The Looking Glass* and in Byron's *Don Juan*, as well as giving the new word 'antimacassar' to the language. It competed as a hair dressing with the long-established bear's grease:

H. LITTLE, Perfumer, No. 1 Portugal Street, Lincoln's Inn Fields, acquaints the Public, that he has killed a remarkable fine RUSSIAN BEAR the fat of which is matured by time to a proper state. He begs leave to solicit their attention to this Animal, which, for its fatness and size, is a real curiosity. He is now selling the fat, cut from the Animal, in boxes at 2s. 6d. and 5s. each, or rendered down in pots, from One Shilling to One Guinea each.

Another advertiser offers any sport-loving gentleman the opportunity of shooting the bear; yet another subtly warns his customers to keep the bear's grease off the backs of their hands, lest they grow hairy paws. As late as 1846 a *Punch* drawing shows a notice in a shop announcing 'Another Bear Just Slaughtered'.

Razors go back to Roman examples of bronze, variously shaped but sometimes so like spearheads as to suggest that the legionary may have sharpened the edges of his spear for this purpose, though Roman troops on active service were not required to shave. In the thirteenth century *rasoors de Guinguant* from Guingamp in Brittany are mentioned as essential to a *lady's* toilet. From the fifteenth century the 'cut-throat' razor is of all tools perhaps the one that has least changed its shape. Louis XI's silver-gilt cut-throats were like those of today. *Boites à savonette* appear in inventories of the seventeenth century; these were silver boxes containing balls of scented shaving soap. A sudden spate of patent razor strops were being advertised in 1710. In 1762 M. Moreau came out with a 'safety razor', a cut-throat type but with a toothed guard, styled a *rabot tricotomique* (or hair-plane). A double-edged razor was offered in 1780. A patent of 1804 covers 'a composition for shaving without the use of razor, soap or water': had any harmless and effective depilatory ever in fact existed, men would surely have given up shaving thenceforth.

AN EASY SHAVE.
LLOYD'S EUXESIS.
For Shaving without Soap, Water, or Brush, and in one-half the ordinary time.
SOOTHING TO THE MOST IRRITABLE SKIN.
Manufactured only by AIMÉE LLOYD, Widow of A. S. Lloyd, 3, Spur Street, Leicester Square, London.
In Metal Tubes, price 1s. 6d. (post free).   Sold by all Chemists and Stores.
Caution.—Ask for the **Widow Lloyd's Euxesis,** and observe "Prepared only by his Widow," in Red Ink across labels.

A CAUTION to prevent IMPOSITION.
SHARP'S CONCAVE RAZORS

Are made of the very best steel that can be possibly procured in this or any other country, tempered, and finished with the greatest nicety and circumspection. Their superior excellence above all others has made them more esteemed than any Razor now in use; the consequence of which is that some persons have offered, and still do offer, an inferior article under their name.

C. SHARP, Perfumer and Razor Maker to his Royal Highness the Prince of Wales, at No. 131, Fleet Street, and No. 57, Cornhill,

Most respectfully intreats the public to observe that his Concave Razors are not sold at any other places in London, but at his shops as above, Sharp stamped on the blade of the Razors; all others are counterfeit.

Sharp's Metallic Razor Strops, which keep the Razor in good order, without the use of a Hone or grinding, are not to be equalled; but the above articles are too well esteemed to need anything being said in their behalf. His Alpine Soap, for shaving, is by far the best adapted for that purpose of any yet invented; it never causes the least smarting sensation, but is perfectly soft, sweet, and pleasing. Likewise his curious Cyprian Wash balls, great variety of shaving cases and pouches, that hold all the implements necessary for shaving, dressing, &c.

Sharp's sweet hard and soft pomatums, are remarkable for keeping good in any climate longer than any other. His Lavender Water, drawn from the flowers, his warranted Tooth brushes and the Prince of Wales Tooth Powder, are articles worthy the attention of the public.

Combs, Soaps, Wash-balls, and every article in the Perfumery branch, wholesale, retail, and for exportation.

N.B. Families, &c, who take any of Sharp's articles by the dozen save considerably.

A complete Dressing case fitted up with razors, Combs &c., for 10s. 6d.

It is disillusioning to learn that when King C. Gillette invented the modern safety-razor (originally costing 5 dollars but soon dropping to 10 cents) his motive was not the improvement of razors. We are told that his actual approach was, 'What can be made that thousands of people will use, that has to be thrown away and replaced as often as possible, but is singly so cheap that the total cost over a period is not noticed?' In pre-Gillette days it was much more usual to be shaved only at a barber's shop. A row of personal shaving mugs bore the names of regular customers. An ingenious but rather frightening device that has disappeared from the barbers' shops was a long revolving pulley-shaft above the row of chairs, driven by steam or electricity. From the pulleys, loose leather belts hung down, and when tightened drove large cylindrical hair brushes with what must have been tremendous effect. English barbers never offered the continental attraction of young ladies who lathered chins (they were not trusted with the razor) but older men in Stockholm can still boast that their chins were once so caressed by a would-be film actress named Garbo.

'Bauer's Head (and Bath) Soap' advertised in the 1880's is said to be also an excellent shaving soap, but it seems a little contradictory that it should also be claimed to promote the growth of hair.

Toothbrushes are of uncertain origin. Roman ladies are said to have used them, and to have chewed mastic from the Isle of Chios to preserve their teeth; even if the teeth were not always in the lady's head, but had to be inserted by a dexterous slave, the mastic was still chewed to keep up appearances. The natives of the African Congo have carried, since times forgotten, willow twigs with frayed ends; the worst insult that can be offered there is 'You haven't cleaned your teeth'. The first known mention of an English 'teeth brush' is in 1651. Queen Elizabeth's teeth, once yellow, were in her old age jet black. A foreign visitor of her time noted the same defect in most of her subjects. The teeth might be rubbed with a linen cloth, or a piece of mallow root, 'to take away the fumosity of the meat and the yellowness of the teeth'. An account of the late sixteenth century includes for 'half a yarde of cloth to rubb my Mrs. teeth, ix *d*.' 'Tooth blanch' and 'tooth soap' are at least as old as this. 'Vaughan's Water' of 1602, according to the maker, 'doth much good to the head and sweeteneth the breath', and is 'better worth than a thousand of their dentifrices'.

C.D.—R

An advertisement of 1660 offers

Most Excellent and Approved DENTIFRICES to scour and cleanse the Teeth, making them white as Ivory, preserves from the Tooth-ach; so that, being constantly used, the parties using it are never troubled with the Toothach; it fastens the Teeth, sweetens the Breath, and preserves the mouth and gums from Cankers and Imposthumes. Made by *Robert Turner*, Gentleman; and the right are onely to be had at *Thomas Rookes*, Stationer, at the Holy Lamb at the East end of St. Paul's Church, near the School, in sealed papers, at 12*d*. the paper.
The Reader is desired to beware of counterfeits.

Lord Chesterfield in 1754 recommends his son to clean his teeth every morning with a sponge and tepid water, with a few drops of 'arquebusade water' added. Soot was a popular dentifrice; recipes for home manufacture used ashes (of nettles or tobacco) mixed with honey; charcoal; areca nuts; cuttlefish bone. Rowland of Macassar Oil fame was making his long-lived 'Odonto' dentifrice, 'prepared solely (*sic*) from Oriental Herbs of inestimable value', before 1800.

The bath mat we have already met, though not yet labelled TAM HTAB, in 1325 when '24 mattis at 2*d*. each' were supplied to the King. Bath mats are shown in some pictures of the mediaeval tub. The bath wrap first appears in 1328 as a *baingnoere*, later *baignoir*, whence the modern garment the *peignoir*. As for towels, we have here another change of meanings, for *toilette* starts as a simple diminutive meaning *petite toile*, a little face-cloth or sometimes the towel draped over the shoulders during shaving. This dates from 1467. An alternative derivation traces the Middle English *towaille* through Old French *touaille* from a Teutonic verb meaning 'to wash'.

The 'flesh brush' was commonly used in the bath until superseded by the sponge, the loofah and the flannel—this last a rather unpleasant material even when new, and itself superseded by the bath-glove made of towelling. The bath-glove was formerly used dry, after the bath-towel, to induce a hearty glow. For those who liked an even fiercer friction, an advertiser of 1895 offered a 'Turkish facial brush of folded horse-hair'. Of sponges, our slight researches reveal nothing noteworthy but the queer fact that they have sex.

Water and steam are by no means the only media that have been used for bathing. Film fans know that ladies such as Poppeia, the wife of Nero, used milk; she never set out on a journey without a train of she-asses to provide it. Beau Brummell also bathed—or let it be known that he bathed—in milk. So did William Douglas, the 4th Duke of Queensberry, or 'Old Q'—a dissolute old sport whose degeneracy was immortalised by both Wordsworth and Burns. It was his habit in old age to espy passing ladies, from his Piccadilly balcony, and to send a valet to pursue and persuade the pretty ones: for all his milk baths he was a dirty old man. He died unmarried, if not childless, in 1810, whereupon many suspicious Londoners resumed the drinking of milk.

Baths have been compounded with seaweed, lees of wine in a state of fermentation, fir-wood oil, camomile, thyme, oak bark, rose petals or walnut leaves; bath-water in England has been as varied a brew as tea in France. A pine-needle bath is less painful than it sounds, and need not be reserved for hardened fakirs, for it employs only a decoction of the needles. Nor need chlorine or hydrochloric acid, which have had their adherents, have been too painful if properly diluted. Compressed air, if used as a medium, should be at a pressure of from 2 to 3 atmospheres; the School of Aviation Medicine at Farnborough would be well equipped to try it with a view to its revival. Among the less inviting materials that have been used are peat ('worked up with water to a pasty consistency'), mud of all kinds, blood (animal or human), horse-dung and guano. Eskimo ladies use urine for washing, but are unlikely to have a supply adequate for bathing. Erasmus says that in the sixteenth century this liquid was popular as a dentifrice. It has, like bees-eggs in the bath-water, been found good for rheumatism. The mustard bath has a long history, from centuries before the 'Mustard Club' advertising campaign of the 1930's. Bran and hot water were the chief ingredients of the 'Arabian Baths' advertised by the beautician Madame Rachel in 1863, at her Bond Street salon, costing 1000 guineas for the full course. Whether a customer would have been found at this figure we shall never know, as the establishment unfortunately had to close when Madame Rachel was sentenced to five years' imprisonment for blackmailing a talkative client. In the naughty nineties the stage star Cora Pearl was said—by her press agent if such then existed—to bathe in champagne. A ladies' magazine of that period advises blondes to wash (though not necessarily to bathe) in hock, and brunettes in claret. It is to be hoped that they did not have access to the keys of their husbands' cellars.

The design of sanitary fittings has for some reason always exercised a special attraction for the 'lunatic fringe' of inventors. Their ideas have ranged from just-practical if odd devices, to the wildest surrealist fantasies. They have been especially prolific in the fields of folding, collapsing and portable gadgets; of the sanitary fitting disguised as something else; of two or more not quite compatible fittings combined into one. Many of these novelties have pleased for a while, but none has lasted.

247

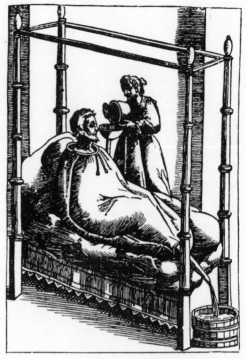

Dr. Sanctorius' Bag Bath

Doctor Sanctorius, who practised in Padua and died in Venice in 1636, invented what we may call the Bag Bath, not unlike the Steam Bag Bath of later times, with its collar that can be tightened round the bather's neck, but intended to hold water. The patient sits in bed, embagged, and the water is poured in through a funnel inserted near the shoulder, to be voided from a tube in the foot of the bag, into a tub. Presumably the outlet can be stopped to give a prolonged soak if required instead of the continuous flow illustrated. The idea might be borrowed today, for a folding pocket-bath for campers: the bag, of waterproof plastic, should allow elbow-room for scrubbing, and a bath taken in two stages (a soapy scrub followed by a clean rinse) should allow complete abstersion with a minimum of water. An assistant is of course needed. If the lower part of the bag were bifurcated, the user need not even be immobilised during the bath, and if arms were formed as well as legs, the assistant might be dispensed with; the idea thus develops of the Pyjama Bath . . . the author expects a share of any profits made from this suggestion.

'Fontaine lavabo' designed by Viollet-le-Duc in 1849. Beaten and engraved copper.
The roof is lifted to fill the reservoir. Two roller towels are provided

249

In 1776 the Comte de Milly invented a *baignoire mécanique*, in which by means of a complicated apparatus the water could be kept moving, to combine the convenience of bathing at home with the fun of bathing in a turbulent river. This idea we have seen reappear over a hundred years later in Ewart's 'Improved Spray Bath' with its 'wave' control.

Sheraton's oddest fancy was an ottoman with 'heating urns' inside, to keep the seat warm in cold weather. Had he been interested in metalwork he might well have gone on from this to produce a self-heating bath, but this luxury was not yet due.

We have already seen 'commodes' disguised as bed-steps or as books, and baths disguised as sofas. Space-economy and convertibility pass the wildest imaginable limits in the 'Piano Bed' of 1866. A real piano unfolds by stages to reveal a writing table, a chest of drawers, a bed, two cupboards for bed-linen, and a wash-basin with jug and towel-rail. The revolving piano stool conceals a lady's work box and a toilet mirror. 'It has been found by actual use that this addition to a piano-forte does not in the least impair its qualities as a musical instrument.'

The Piano Bed, 1866

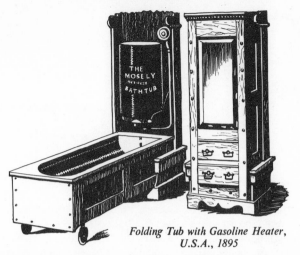

Folding Tub with Gasoline Heater,
U.S.A., 1895

Some of the folding and combination sanitary furniture has been fairly sensible. The American 'Wardrobe Bath Tub' of 1895, with a 'gasoline heater', was a good answer to the space problem; the Montgomery Ward mail order catalogue in which it appears offers more folding tubs than normal ones. McDowall, Stevens & Co's 'Patent Bath' of 1897 has its fittings all below the rim of the bath, and a hinged flap covers the whole, 'forming a lounge or bedstead, where room is of consequence'.

Around 1900 comes a whole series of essays in compressibility. Jennings manages to combine a bidet, a footbath, a Sitz bath and a 'commode pail' in one reversible double receptacle, that must have needed cautious handling. This recalls the two-purpose basin used by the Duc de Vendome, *qui, sortant tout chaud de la chaise percée et à peine lavé, lui servait de suite de bassin à barbe.*

Reversible Commode and Bidet

Swinging Washbasin, 1907

Adamsez come out with a series of sanitary conjuring tricks. A swinging wash-basin swivels over the bath, into which its waste falls, and with which it presumably shares the taps. The 'Patent Concealed Closet Set' has a cylindrical water-closet over which a cylindrical wash-basin nests neatly, or swings aside. Another portmanteau series culminates in an arrangement whereby the water-closet, its overhead tank, the wash-basin, the medicine cabinet and the mirror all occupy the same plan space. Of some of these devices it must be remarked, that if there was space to swing the wash-basin as shown, this space might surely have been used for a fixed basin, and a lot of hinges and nuts and bolts saved.

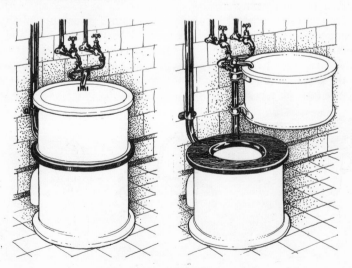

Patent Concealed Closet Set, 1907

Swing Lavatory 1907

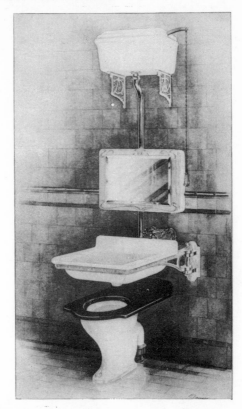

Solid Copper Hospital Bath

" LAURA "

Improved Russian Bath or Combined Vapour and Hot Air Bath, 1890: The hot pipes prevent the steam from condensing

Portable Turkish or Hot Air Bath, 1899

Glamour and the Bath: French, German and English

Shanks, in their 'Railway and Ships Catalogue', show many justifiably compact fittings, and among them one unexpected luxury: a rack to hold bottles, for the railway lavatory compartment, a brandy bottle being shown in the illustration. This fitting may have been intended only for private compartments.

The Jeep Bath, popular in hot and waterless places during World War II, was sane and sensible as long as it consisted only of a metal Jeep trailer filled with precious and far-borne water heated under a tropical sun, and then used in descending order of rank until repulsion set in at about Flight-Lieutenant level. It was not until a humorist conceived the idea of starting up the attached Jeep and touring the helpless occupant of the trailer round the district, that this type of bath qualified as a fit conclusion to this digressive chapter of miscellaneous eccentricities.

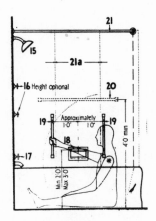

XVIII

Millions of Baths

*New Wine In Old Bottles — The London House — Improvised
Bathrooms — Space-saving — Percolation of Styles — Grandfather Not Surprised — Agonised Wife's Dilemma — The Real-
life Bathroom — Black Marks For Plumbing — Sour Criticisms
— Our Habits — New Devices — Conclusion*

S OON after the first World War enormous numbers of 'enamelled sanitary fixtures' (mainly baths and wash-basins) were being turned out: in the two years from 1921 in the U.S.A. the annual number rose from 2,400,000 to 4,800,000. In Great Britain there was a similar increase, but these millions of new baths by no means all found their way into the sort of bathroom that the catalogues of the period depict. New wine usually had to go into old bottles. The average age of the English middle-class houses was probably at least thirty years. Few houses of this age before the 1920's would have a bathroom that had been built as such. The standard London terrace house of the better class, built about the 1860's, had gone up in thousands on an almost invariable plan, with only two rooms of any importance on each of its four or five floors. It was meant for the growing prosperous middle or upper-middle class family, with at least one servant to toil up and down its extravagant stair-well at the call of a bell, carrying buckets of coal, cans of water and trays of food. In 1850 domestic workers formed the largest 'trade group' in London. Only six English cities had a larger *total* population than the 121,000 'general domestic servants' of London. Well over half of these were aged under twenty-five. The housemaid suffered some of the pains of a lighthouse-keeper. The basement was dark and airless. The attics were angular and low, barely 'habitable rooms within the meaning of the Act', and tanks hissed and gurgled there. It was a stupid house plan; the space was adequate but ill-organised. Such houses had a short life in their intended role as family dwellings. Within a generation, or two at most, they were proving unworkable. The housemaid had found in the munition factory comparatively easy work and far better pay, and did not return in 1918. Sooner or later the house was broken up into 'maisonettes',

258

flats of a sort, or single furnished rooms whose occupants passed furtively on the stairs to rattle at the locked door of the bathroom or the water-closet. Rows of such houses were converted into rambling private hotels with a superfluity of unused classical porticoes.

Before such rearrangements, many households still made shift with portable baths. Except for the larger suburban houses, those that installed a bathroom could seldom spare a whole large bedroom for conversion, and they improvised in various ways. A bathroom could be added above the single- or two-storey 'offices' at the rear: rows of oddly-assorted bathrooms so added may be seen from the back gardens of Bayswater, Pimlico and Kensington, their individualistic architecture and plumbing intensifying the effect of 'Queen Anne fronts and Mary Ann backs'. A bathroom could be filched from a bedroom if one was not too sensitive, lying in bed or bath, to the effect on the ceiling mouldings; curious things happened when partitions met windows. Stair landings were sometimes just big enough to squeeze in a bath cubicle, through which borrowed light and watery noises penetrated to the stairway.

When the house was chopped into flats, and more baths had to be fitted in, the kitchen—itself improvised—offered a site where the bath could share the geyser. A bathroom for the basement flat would be contrived in the 'glory hole' under the entrance steps, where a minimum of light and air combined with a maximum of piping and meters. In these cramped quarters, folding baths, though still sold in some variety, were less popular than one would have expected (perhaps because they were more expensive than the now very cheap standard tub) but much ingenuity still went into space-saving.

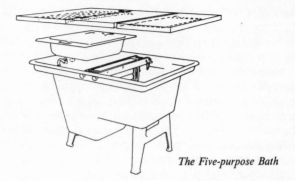

The Five-purpose Bath

The '5-purpose Bath' of the International Bath Association combines a strong table for general use, a kitchen sink with a draining-board, a wash-basin, a 'home laundry' with a wringer, and a bath 'for adult or child'. There is, however, no shower or water-closet. Another space-saving bath from America requires the user to take an attitude reminiscent of a crouch-burial.

More recently, Buckminster Fuller's 'Bathroom-Kitchen-Heat-Light Unit' combines in one prefabricated mechanical core almost everything but the living room and bedrooms. Space-saving can go no further. The resulting little 'machine for living in' is perhaps not quite so acceptable for life in a decent home, as it might be for a brief sortie in a submarine. It would be a pity if space- and cost-saving should end by giving the house all the disadvantages and none of the advantages of the caravan.

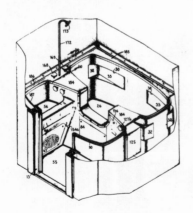

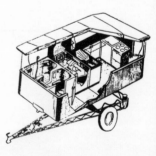

Buckminster Fuller's Prefabri-cated Bathroom, 1938, and his Mechanical Core, 1943

Not all the new baths went into the improvised quarters of the New Poor. Council houses and flats have provided millions of baths in neat new little bathrooms, enough for nearly all of those who can not afford them. Nothing technically very noteworthy has arisen in the process. For a long time bathrooms have varied little, except for finish and decorative style. Edwardian Renaissance and *art nouveau* having passed in their turn, we now see the last unhappy throes of the *style moderne* that slowly percolated down in the years following the Paris Exhibition of 1925—as all styles do—until it reached the 'spec builder's' level. But for reflecting these passing fashions, the bathroom has become standardised. We have little to show there that would surprise our grandfather. One manufacturer currently lists nineteen different shapes or sizes of baths, and a choice of seventy-two colours, of which pink is available in twenty-two shades. His showroom, if every variant is displayed, will contain 1,368 baths, but the smallest of these is only the standard model scaled down; they are essentially all the same. Our grandfather had a far wider choice. His almost indispensable shower has become a rarity in England, where the American glazed 'tub screen' is seldom seen. (Perhaps we have been deterred by that terrible scene depicted by Peter Arno: the door of the glass enclosure, and the shower-tap, have both jammed, and the gesticulating husband is submerged inside. Worse still is the dilemma of the agonised wife outside—what will happen if she opens the door?)

Let us consider the bathroom of today—as it really is. Let the reader go now to his bathroom and look at it dispassionately for a few minutes. In what proportion of bathrooms has the problem of the junction between a rounded bath and a straight wall been neatly and lastingly solved? We have abandoned that too-ingenious form of bath-waste operated by twisting the top of a heavy vertical tube; it was hard to clean and seldom watertight; where these survive, how often do we not arouse from steamy idleness to find ourselves half stranded? We have reverted to the plug and chain—but what proportion of these chains, in real life, are intact and stringless? An alternative to visible taps is offered in a leading London hotel, where pre-mixed bathwater is delivered, at a terrifying rate, up through the combined inlet and waste-outlet, invariably bringing with it the last two or three ounces of the previous user's residue; let others say whether the telephone installed in the adjoining water-closet is adequate compensation. How many wash-basins have an overflow that is cleanable? Who has dared to experiment there with a flexible probe? What of the condensation problem? The lighting of the mirror? The heating of the bathroom and of the water-closet? What

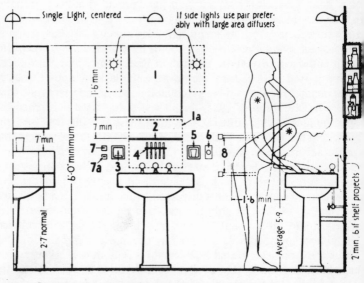

Basic Dimensions for the Bathroom (American Architect, *1935*)

of that medicine cabinet and its contents—a history of the family's ailments over the past decade? How many bathrooms are put out of action by a few days of frost? How many have that accessory the 'Plumber's Delight'? (This should not be confused with 'Builder's Delight', which is a particular shade of hot brown beloved of house-painters. It is the informal trade name for a glass shelf above the wash-basin, carried on brackets set too far apart, so that when dis-placed by a few inches it falls into the basin with a good chance of cracking it; this is good for trade, hence the name.) As for untidy plumbing: if we assess our bathroom in this respect on a points system, giving one black mark for each visible joint, bend or pipe-clip (water or gas), how many will score less than twenty? (The author's quite neat bathroom, with cased bath and basin, scores twenty-six, which may be taken as bogy for future contests.)

How many real-life water-closets work at the first pull (practised family hands excepted), and do their job every time? And at what decibel level? How many doctors, even if they have not travelled in the Orient, will agree that the closet seat is at the right height, and the user's attitude natural? How many agree that the 'Eastern Pedes-tal Closet' with its squatting position is, as one writer has it, 'only suitable for native races'? (What is a 'native race'?)

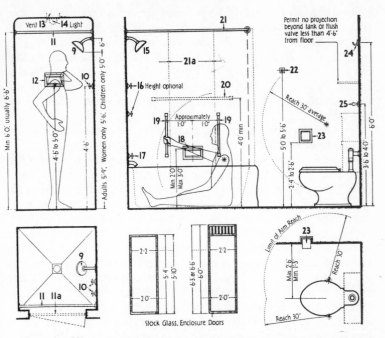

A. J. Lamb has quoted a report of the 1930's to the effect that at least 10,000 persons were injured annually in bathrooms, by slipping, by gas poisoning or by electrocution. Neither plumber nor householder, neither architect nor manufacturer, can yet sit back in complacent contemplation of a problem solved. Of these, only the householder need object to our sour and negative criticisms, which mean plenty of work for the other three in the years ahead. There are, of course, impeccable bathrooms, but the *average* bathroom, in relation to the technical means available and our skill in far more complex installations, is hardly worthy of the *sputnik* age. Perhaps our equipment is what we can afford, and we do mean to do something soon about that bathroom: a new water-closet would cost less than a television set. But the 1958 edition of *Television Factbook* reveals that in the U.S.A., 42,400,000 or 84 per cent of the nation's homes have one or more television sets, whereas nearly a million fewer have bathtubs. According to the West German Baths Society, inspectors who visited 200 homes in a west German town in 1958 saw 125 television sets but only three baths.

A housing survey in 1958 in Morley, Yorkshire, where one house in five was scheduled for clearance, showed that 50 per cent of the houses had no baths and 30 per cent shared a lavatory. Another rather vaguely-worded estimate tells us that 'about one third of the old dwellings' in Great Britain are bathless; *The Times* tells us (6/3/58) that in the Gorbals in Glasgow less than one house in four has a 'lavatory', and about one in 30 a bath—the same situation as existed 30 years ago. Not long since, 'theatrical digs' in Manchester were advertised as enjoying advantages neatly abbreviated as 'lav.in; pub opp.'

So much for our equipment; accepting it, what meanwhile of our personal habits? Will posterity rate us clean and decent?

Although the modern bath offers a rather superficial cleansing as compared with the Islamic method, it is surely adequate if we use it daily? Perhaps we have never quite recovered from the habits of the 'peak period' days and the 5-in. Plimsoll line of the wartime bath-water ration. But even if we bathe but weekly, we should stand fairly high in history. We may or may not deflate when a statistician tells us that of our neighbours on a London bus today, one in five *never* takes a bath. Many of our cinema managers find it necessary to spray their audiences with *sparsiones* such as once sweetened the Roman rabblement. We do not cry 'Gardy Loo!', but the wynds of mediaeval Edinburgh and the field paths where Cobbett rode were not ankle-deep in waste paper. When the legions marched along Icknield Way they did not wonder, as we do, to see the gullies of Dunstable Downs white with a summer snow of ice-cream cartons. Mediaeval dogs were no nastier in their street behaviour than ours. Erasmus would see in our gutters *sordes* of a new kind that even to us are *non nominandas*. In India it is a sound sanitary custom not to let the right hand know what the left hand doeth, and in the monasteries the proper route from the rere-dorter to the refectory was by way of the laver. But a recent campaign to enforce this custom, particularly in catering establishments, did not find universal support: its advertisement was refused by a Top People's daily paper which, when pressed, replied coldly that all of *its* readers already observed the rule.

But we had agreed not to dwell on things of the present day. The advertisements speak very highly of the thermostatic mixing tap, the spray tap, copper and plastic pipes, the Perspex bath, the invisible towel of hot air, and the new electric razor that relieves the user from the effort and time-lag of switching it on—for a few dollars extra, he can save daily seconds that accumulate over the years, for this model starts when lifted from its hook. Let the future historian evaluate such achievements, and rewrite this last chapter.

SHORT BIBLIOGRAPHY

ASHE, G. *The Tale of the Tub* (Newman Neame; London, 1950).

BELL, J. *A Treatise on Baths* (Barrington; Philadelphia, 1850).

BESANT, SIR W. *The Survey of London: London, City* (Black; London, 1910).

BUER, M. C. *Health, Wealth and Population in the Early Days of the Industrial Revolution* (Routledge; London, 1926).

CHEYNE, G. *An Essay on Health and Long Life* (Strahan; London, 1724).

CHIPPENDALE, T. *The Gentleman and Cabinet Maker's Director* (London, 1754).

CROSSLEY, F. H. *The English Abbey* (Batsford; London, 1935).

EASSIE, W. *Sanitary Arrangements for Dwellings* (Smith, Elder; London, 1874).

EVANS, Sir A. *The Palace of Minos* (Macmillan; London, 1921).

FIENNES, C. See MORRIS, C.

FOURNIER, E. *Le Vieux Neuf* (Dentu; Paris, 1859).

FREEMAN, H. W. *The Thermal Baths of Bath* (Hamilton Adams; London, 1888).

FUCHS, E. *Illustrierte Sittengeschichte* (Langen; Munich, 1909).

GIEDION, S. *Mechanisation Takes Command* (Oxford University Press; New York, 1948).

GLOAG, J. *Georgian Grace* (Black; London, 1956).

GRANVILLE, A. B. *The Spas of England* (Colburn; London, 1841).

HAVARD, H. *Dictionnaire de l'Ameublement* (Paris, 1890–4).

HELLYER, B. *Under Eight Reigns* (Dent & Hellyer; London, undated; *c.* 1930).

HELLYER, S. S. *The Plumber and Sanitary Houses* (Batsford; London, 1877). *Principles and Practice of Plumbing* (Bell; London, 1891).

HEPPLEWHITE, G. *Cabinet Maker's and Upholsterer's Guide* (London, 1787).

LANE, R. J. *Life at the Water Cure* (Longman; London, 1846).

LLOYD, N. *A History of the English House* (Architectural Press; London, 1930).

LUCAS, C. *An Essay on Waters* (London, 1756).

MAYHEW, H. See QUENNELL, P.

MOORE, E. C. S., AND SILCOCK, E. J., *Sanitary Engineering* (London, 1901).

MORRIS, C. (Ed.) *The Journeys of Celia Fiennes* (Cresset Press; London, 1947).

PUDNEY, J. *The Smallest Room* (Joseph; London, 1954).

QUENNELL, P. *Mayhew's London* (Pilot Press; London, 1949). *Mayhew's Characters* (Kimber; London, 1951).

REYNOLDS, R. *Cleanliness and Godliness* (Allen & Unwin; London, 1943).

ROBINS, F. W. *The Story of Water Supply* (Oxford University Press, 1946).

SALZMAN, L. F. *Building in England Down to 1540.* (Oxford University Press, 1952).

English Life in the Middle Ages (Oxford University Press, 1926).

SAMPSON, H. *A History of Advertising* (Chatto & Windus; London, 1874).

SANDSTRÖM AND THUNSTRÖM. *Bad I Hemmet Förr Och Nu* (Haeggström; Stockholm, 1946).

SCOTT, G. R. *The Story of Baths and Bathing* (T. Werner Laurie; London, 1939).
SHEARER, T. *The Cabinet Maker's London Book of Prices* (London, 1788). *Designs for Household Furniture* (London, 1788).
SHERATON, T. *The Cabinet Dictionary* (London, 1803).
SINGER, HOLMYARD, HALL and WILLIAMS. (Ed.). *A History of Technology* (Oxford University Press; 1956).
SITWELL, O., and BARTON, M. *Brighton* (Faber & Faber; London, 1935).
TEALE, T. P. *Dangers to Health* (Churchill; London, 1878).
TREVELYAN, G. M. *English Social History* (Longmans, Green; London, 1944).
TUER, A. W. *Luxurious Bathing* (Field & Tuer; London, 1880).
TURNER, E. S. *The Shocking History of Advertising* (Joseph; London, 1952).
VALLENTIN, A. *Leonardo da Vinci* (Allen; London, 1952).
VIOLLET-LE-DUC, E. E. *Dictionnaire Raisonné du Mobilier Français* (Morel; Paris, 1871).
WARD, J. *Romano-British Buildings and Earthworks* (Methuen; London, 1911).
WARE, I. *A Complete Body of Architecture* (London, 1756).
WELLS, R. B. D. *Good Health, and How to Secure It* (Vickers; 1885).
WILLIAMS, N. *Powder and Paint* (Longmans, Green; London, 1957).
WRIGHT, T. *A History of Domestic Manners and Sentiments in England During the Middle Ages* (Chapman & Hall; London, 1862).
YARWOOD, D. *The English Home* (Batsford; London, 1956).
YOUNG, G. M. (Ed.) *Early Victorian England, 1830–1865* (Oxford University Press; 1934).

PERIODICALS, PAPERS, etc.

CAMERON, R. *Salles de Bains* (*l'Oeil*, Paris, May 1958).
Gas Journal, Centenary Number, 1949.
HOPE, W. H. St. J. *The London Charterhouse and its Water Supply* (*Archaeologia*, 58, Part I).
LAMB, H. A. J. *Sanitation: An Historical Survey* (*The Architects' Journal*, 4/3/37).
SABINE, E. L. *Latrines and Cesspools of Mediaeval London* (*Speculum*, U.S.A., July 1934).
SYMONDS, R. W. *The Craft of the Coffermaker* (*The Connoisseur*, March 1941).
THORESBY SOCIETY: *Kirkstall Abbey Excavations*, 2nd report, 1951; 6th report, 1955 (Thoresby Society, Leeds).
WEBSTER, D.McK. *Enamels and Enamelling, with Special Reference to Baths* (unpublished paper).

TRADE CATALOGUES, etc.

Adamsez Ltd.; Allied Ironfounders Ltd.; John Bolding & Sons Ltd.; Bostel Bros. Ltd.; The British Bath Co. Ltd.; Dent & Hellyer Ltd.; Doulton & Co. Ltd.; Ewart & Son Ltd.; W. N. Froy & Sons Ltd.; George Jennings Esq.; E. Johns & Co. Ltd.; Rownson, Drew & Clydesdale Ltd.; Shanks & Co. Ltd.; Twyfords Ltd.; Chas. Winn & Co. Ltd.

ILLUSTRATIONS: SOURCES AND ACKNOWLEDGEMENTS

(Works included in the Bibliography are identified here by the authors' surnames only and illustrations are identified by their page numbers)

Page iii, Fuchs, from fifteenth century woodcut. 1, Singer, from Greek vase, sixth century B.C. 5 (i), Adams (1896). 5 (ii), 6 (i), 6 (ii), Redrawn from Evans. 7, Evans. 8 (i), 8 (ii), Redrawn from Evans. 9, Evans. 10, Redrawn from Singer. 11 (i), Scott, from Wilkinson's *Ancient Egyptians*; (ii), Redrawn from photograph by Egypt Exploration Society. 13, Smith, *Dictionary of Greek and Roman Antiquities*. 15, Freeman. 16, Redrawn from *Encyclopaedia Britannica*, 11th ed. 19, Leaflet published by Corporation of Bath. 20, Redrawn from Ward, from O. Morgan. 23, Hellyer, B., from drawing by F. R. Dickinson. 25, Redrawn from a restoration by A. C. Henderson, F.S.A. 26, Singer, from Psalter of Eadwin, c. 1167, at Trinity College, Cambridge. 27, Redrawn from Salzman from source as above. 28, Redrawn from London Charterhouse Library. 30, Drawn from a photograph. 31, Williams, *History of the G.W.R.* 33, Fuchs. 35, Havard, from *Chronique et Histoire des Quatre Monarchies du Monde*, Bibliothèque de l'Arsénal, Paris. 37, Viollet-le-Duc. 38, Dürer, *Life of the Virgin*, 1509. 39, Wright, T. 40 (i), Havard, from *Aesop's Fables*, 1501; (ii), Fuchs. 41, Wright, T. from thirteenth-century MS. 43 (i), Sandström; (ii) Fuchs. 44 (i), Havard, from *L'Histoire de Jason*; (ii) Havard, from fifteenth century woodcut; (iii), Havard, from Bibliothèque de l'Arsénal, Paris. 45 (i), Havard, from tooled leather casket, fifteenth century; (ii), Dürer, *Life of the Virgin*, 1509; (iii), Havard, from *Roman de Watriquez*. 46, 47, 48, Hellyer, B., from drawings by F. R. Dickinson. 50, Detail from Visscher's *View of London*. 51, Havard, from *The Decameron*, fifteenth century, Bibliothèque de l'Arsénal, Paris. 52, Singer. 53, Wright, T., from Harleian MS. No. 603. 55, Sandström, from woodcut by Hans Beham, sixteenth century. 58, Sandström. 59, 60, Dürer. 61 (i) Fuchs; (ii), Sandström. 62 (i), (ii), Fuchs, from woodcuts by Jost Amann. 64, Robins, from *Old England*. 67, Fuchs, from engraving by J. M. Will, Nuremberg. 69, 70, Drawn from photographs. 72 (i), (ii), 74, Science Museum Library, from Harington. 75, Drawn from portrait in the National Portrait Gallery. 77, Hogarth, *Four Times of the Day*, 1738. 79, Fuchs. 80, Redrawn from a drawing by Thomas Johnson in the British Museum. 82, Rowlandson. 84, 85, (i), (ii), Fuchs. 87 (i), Scott, from *The Builder*, 1861; (ii), Scott, from *Archaeologia*, 1834. 89, Fuchs, from a woodcut by Jost Amann. 91, Besant. 93, Drawn from a photograph from Sandström. 94, Besant. 97, Redrawn from Hellyer, S. S. 99, Fuchs, from a print published by R. Sayer, London. 100, 101, Fuchs, from an engraving by Theodor de Bry, from a woodcut by H. S. Beham. 102, Havard. 105, Ware. 106, 107, 108, Hellyer, S. S. 109, Rowlandson. 111, Sheraton. 113 (i), Havard; (ii), Chippendale; (iii), Havard; (iv), Sheraton. 114, Hepplewhite. 116 (i), (ii), Chippendale.

116 (iii), 117 (i), (ii), **Hepplewhite.** 117 (iii), Sheraton. 118 (i), (ii), Havard. 119, Viollet-le-Duc. 120, **Hepplewhite.** 121 (i), Havard, from *Aesop's Fables*, 1501; (ii), Fuchs, from German woodcut, fifteenth century; (iii), Havard, from a picture by Brekelenkam, The Louvre, Paris. 125, Havard, from Boucher. 127 (i), (ii), Havard. 128 (i), Fuchs, from engraving by Crispin de Passe, from picture by Martin de Vos; (ii), Fuchs, from engraving by Georgius, from mural painting by Raphael. 129, Fuchs, from engraving by Le Beau, 1773. 130, 131, Drawn from photographs from *l'Oeil*. 132, Fuchs, from engraving by Girard from *A Day in the Life of a Courtesan*. 133, Drawn from a photograph from *l'Oeil*. 134, *l'Oeil*. 135 (i), Havard; (ii) Havard, from Percier and Fontaine. 136, Redrawn from Havard. 137, Fuchs, from an engraving by L. Surugue, from a picture by L. B. Pater, 1741. 141, Quennell, from Mayhew. 143, Hellyer, S. S. 145, London Museum. 146 (i), (ii), L.C.C., *The Centenary of London's Main Drainage*. 147, Williams, *History of the G.W.R.* 150, Flybill of 1832. 152, 153, Doré, *London*, 1877. 154, 155, Quennell, from Mayhew. 157, Lane. 159, Giedion, from Fleury, *Traité Hydrothérapique*. 160, 161 (i), (ii), Giedion. 162, Ewart. 163, Turner. 167 (i), (ii), (iii), (iv), 168 (i), (ii), Ewart. 169, Wells. 170 (i), (ii), 171 (i), Ewart. 171 (ii), Wells. 172 (i), (ii), Ewart. 173, Havard. 174 (i) Redrawn from Jennings; (ii), (iii), 175 (i), (ii), Ewart. 177, Lane. 181 (i), Scott, from Urquhart, *Manual of the Turkish Bath*; (ii) Giedion. 184, Wells. 187, Redrawn from Jennings. 189, Lane. 190 (i), Giedion, from Cole, *Journal of Design*, 1850; (ii) Redrawn from Jennings. 192, Giedion, from plumber's advertisement, Boston, 1850. 193, Ewart. 194 (i), (ii), Redrawn from *Gas Journal*. 195 (i), Froy; (ii), Redrawn from Ewart; (iii), Ewart. 196, Froy. 197, Ewart. 199, Hellyer, S. S. 201 (i), Lamb; (ii), Adams. 202 (i), (ii), (iii), Redrawn from Lamb. 203 (i), (ii), (iii), Dent and Hellyer. 203 (iv), 204, Jennings. 205, Redrawn from Adams. 206 (i), (ii), Doulton; (iii) Anonymous catalogue lent by E. Johns & Co., Ltd.; (iv), Doulton; (v), Shanks. 207 (i), Bolding; (ii), Doulton; (iii), Froy. 209 (i), (ii), (iii), Bolding. 211, Hellyer, S. S. 212, 213, Teale. 214 (i), Building Trades Exhibition catalogue, 1894; (ii), Hellyer, S. S.; (iii), Building Trades Exhibition catalogue. 215, Dent and Hellyer. 217, Hellyer, S. S. 219 (i), Drawing by the author; (ii) Hellyer, S. S.; (iii), (iv), Bolding. 220 (i), Hellyer, S. S.; (ii), Shanks; (iii), Hellyer, S. S. 221 (i), Doulton; (ii) Hellyer, S. S.; (iii) Twyford; (iv), Doulton. 223 (i), (ii), Froy; (iii), Doulton; (iv), Shanks. 224 (i), Redrawn from Giedion, from catalogue 1885; (ii), Lamb, from Jennings. 225 (i), Shanks; (ii), Doulton. 226 (i), Shanks; (ii) Ewart; (iii), Froy. 227 (i), (ii), Jennings. 229 (i), Ewart; (ii), Ewart; (iii), Froy. 230, 231, Shanks. 232 (i), Froy; (ii) Ewart; (iii), Adamsez. 233 (i), Shanks; (ii), Adamsez. 234, Giedion. 235 (i), (ii), Shanks; (iii), Redrawn from Giedion. 236 (i), (ii), Froy; (iii), Redrawn from Shanks; (iv), Shanks. 237 (i), (ii), Adamsez. 239, Jennings. 241, Fuchs, from an engraving by Le Blond. 242, Advertisement for Pears Ltd. by Ernest Griset, 1898. 244, Sampson, from *The Times and Daily Register, 1788*. 248, *l'Oeil*. 249, Viollet-le-Duc. 250, Giedion. 251 (i), Redrawn from Gieidon, from Montgomery Ward Co. catalogue; (ii), Jennings. 252 (i), (ii), Redrawn from Adamsez. 253 (i), Adamsez; (ii), Redrawn from Ewart. 254 (i), (ii), (iii), Shanks. 255 (i), (ii), Fuchs; (iii), *Punch*, 1891, by Du Maurier. 256, From a photograph by the author, Tunisia, 1943. 257, *American Architect*, 1935. 259, Building Trades Exhibition catalogue, 1905. 260 (i), Lamb, from International Bath Association; (ii), Giedion, from Bruce and Sandbank, *A History of Prefabrication*, N.Y., 1944. 262, 263, *American Architect*, 1935. 265, Fuchs. 282, Redrawn from Bolding.

INDEX

271

INDEX

273

Mari 10
Marly 99
Marseillaise 101
Mary, Queen of James II 81
Marylebone 62
Mass-production 238, 258
Mats (*see* Bath mats)
Maughan, B. W. 193, 194
Mayhew 139, 149, 151, 153–6
Mazarin, Cardinal 122
McDowall, Stevens 251
Mechanical core 260
Medical baths (*see* Hydropathy)
Medicinal waters (*see* Spas)
Meissen 122
Meistersingers 59
Menelaus 12, 36
Metal baths (*see* Copper, Lead, Iron, Tin)
Metamorphosis of Ajax 71–4
Metropolitan Water Board 65
Metternich, Prince 124
Michael Angelo 14
Midland Railway 142
Milford Haven 57
Milk baths 246, 247
Milly, Comte de 250
Mineral waters (*see* Spas)
Minerva 80
Minos, King 4
Minton 124
Mixing tap 193
Monasteries (Abbeys, Priories) 2, 24–32, 75, 264; Beaulieu 28; Bury St. Edmunds 28; Charterhouse, London 28, 29; Chester 28; Christ Church, Canterbury 25–27, 29, 31; Churchdown 28; Dublin 63; Durham 25, 29, 31; Fountains 24, 31; Furness 31; Gloucester 25, 28; Kirkstall 24, 29, 31; Lacock 28; Lewes 31; Much Wenlock 25; Norwich 25; Reading 28; Redburn 32; Sherborne 28; Tintern 32; Westminster 144
Montague, Lady Mary Wortley 88
Montgomery, Mrs. M. A. v, ix
Moorish bath (*see* Turkish)
Moreton closet enclosure 202, 203
Morice, Peter 65
Morley 263
Moses 10
Mottram, R. H. 144

Moule, Rev. H. 208
Mount Vernon Hotel 234
Much Wenlock Priory 25
Mud baths 164, 247
Muffet, Dr. T. 140
Multi-point geysers 198
Murders in the bath 80, 172, 173
Musical chamber pot, night table 124
Mustard bath 247
Myddleton, Sir Hugh 65
Mystères de Paris 102

Nala, King 56
Napoleon I 123, 130
Nash 83
Natron 240
Nausicaa 21
Necessarium, Necessary house 47
Nekt 11
Neolithic latrines 4
Neva, River 57
Newcastle 34
Newgate Jail 49
New River 65, 93, 149
New York 118, 165, 224
Niagara Falls 159
Night men 145, 147, 156
Night table 115, 117, 118, 124
Nile, River 10
Norfolk, 11th Duke of 138
Norwich 25, 137
Nose-blowing 25, 36
Nottingham 144
Nuisances Removal Act 151
Nuremberg 38, 59

Odonto Dentrifice 246
Odyssey 12, 21, 36
Office houses 103, 104
Optimus closet 202, 203
Order of the Bath 57
Ordinances; Plumbers' 66
— Sanitary and Housing 50, 52, 60, 90, 94
Orignal 120–2
Orkneys 4
Osborne 227
Osterley House 104
Otazel heater 193
Oxford 2, 76

Padding in baths 42, 57, 98, 126, 127
Paddington 63, 155